Uniform Laws and Regulations in the Areas of Legal Metrology and Engine Fuel Quality

as adopted by the 98th National Conference on Weights and Measures 2013

Editors:
Linda Crown
David Sefcik
Lisa Warfield

Carol Hockert, Chief
Office of Weights and Measures
Physical Measurement Laboratory

U.S. Department of Commerce
Penny Pritzker, Secretary

National Institute of Standards and Technology
Patrick D. Gallagher, Under Secretary of Commerce for Standards and Technology and Director

NIST Handbook 130
2014 Edition

Supersedes NIST Handbook 130, 2013 Edition

National Institute of Standards and Technology Handbook 130, 2014 Edition
Natl. Inst. Stand. Technol. Handb. 130, 2014 Ed., 300 pages (October 2013)
CODEN: NIHAE2

WASHINGTON: 2013

Foreword

This handbook compiles the latest Uniform Laws and Regulations and related interpretations and guidelines adopted by the National Conference on Weights and Measures, Inc. (NCWM). At the 1983 Annual Meeting, the NCWM voted to change the title of Handbook 130 and the title of the Laws and Regulations compiled in this handbook. The former title of the handbook was "Model State Laws and Regulations." "Model State" was to be changed to "Uniform" in the title to reflect that these Laws and Regulations are (a) intended to be standards rather than just guidelines, and (b) intended for adoption by political subdivisions other than states when deemed appropriate.

This edition includes amendments approved at the 98th NCWM Annual Meetings in 2013. The NCWM recommends adoption and promulgation by weights and measures jurisdictions of these Uniform Laws and Regulations as updated in this handbook.

The National Institute of Standards and Technology (NIST) has the statutory responsibility to promote "cooperate with the states in securing uniformity in weights and measures laws and methods of inspection." In partial fulfillment of this responsibility, the Institute is pleased to publish these recommendations of the NCWM.

This handbook promotes the primary use of the International System of Units (SI) by citing SI units before customary units where both units appear together, and by placing separate sections containing requirements for SI units before corresponding sections containing requirements for customary units. In some cases, however, trade practice is currently restricted to the use of customary units; therefore, some requirements in this handbook will continue to specify only customary units until the NCWM achieves a broad consensus on the permitted metric units.

THIS PAGE INTENTIONALLY LEFT BLANK

Committee Members

Committee on Laws and Regulations of the 98th Conference

Judy Cardin, Wisconsin
*Raymond Johnson, New Mexico
Richard Lewis, Georgia
Tim Lloyd, Montana
Louis Sakin, Towns of Hopkinton/Northbridge, Massachusetts

Associate Membership Committee Representative: Rob Underwood, Petroleum Marketers Association of America
Fuels and Lubricants Subcommittee: Ron Hayes, Missouri
Packaging and Labeling Subcommittee: Christopher Guay, Procter and Gamble
Canadian Technical Advisor: Lance Robertson, Measurement Canada
NIST Technical Advisors: Lisa Warfield, David Sefcik
NIST Technical Advisor on the Uniform Regulation for National Type Evaluation: Tina Butcher
*Acting Chair for the NCWM Annual Meeting

Past Chairmen of the Committee

Conference	Chairman
41	G. H. Leithauser, MD
42	F. M. Greene, CT
43	F. M. Greene, CT
44	G. L. Johnson, KY
45	R. Williams, NY
46	J. H. Lewis, WA
47	J. H. Lewis, WA
48	J. H. Lewis, WA
49	J. H. Lewis, WA
50	L. Barker, WV
51	L. Barker, WV
52	M. Jennings, TN
53	W. A. Kerlin, CA
54	J. F. Lyles, VA
55	J. F. Lyles, VA
56	S. D. Andrews, FL
57	S. D. Andrews, FL
58	S. D. Andrews, FL
59	R. M. Leach, MI
60	R. L. Thompson, MD
61	C. H. Vincent, Dallas, TX
62	C. H. Vincent, Dallas, TX
63	J. T. Bennett, CT
64	R. W. Probst, WI
65	D. I. Offner, MO
66	J. J. Bartfai, NY
67	J. J. Bartfai, NY
68	J. J. Bartfai, NY
69	W. R. Mossberg, CA
70	E. Skluzacek, MN
71	D. Stagg, AL
72	A. Nelson, CT
73	K. Simila, OR
74	K. Simila, OR
75	S. B. Colbrook, IL
76	A. Nelson, CT
77	B. Bloch, CA
78	F. Clem, OH
79	B. Bloch, CA
80	S. Rhoades, AZ
81	L. Straub, MD
82	S. Millay, ME
83	K. Angell, WV
84	K. Angell, WV
85	S. Morrison, CA
86	R. Williams, TN
87	P. D'Errico, NJ
88	D. Johannes, CA
89	D. Johannes, CA
90	J. Gomez, NM
91	J. Benavides, TX
92	J. Cassidy, MA
93	Vicky Dempsey, OH
94	Joe Gomez, NM
95	Joe Benavides, TX
96	John Gaccione, NY
97	Judy Cardin, WI

Table of Contents

2013 Amendments

The following table lists the laws and regulations amended by the 98th (2013) National Conference on Weights and Measures (NCWM). As appropriate, the text on the cited pages indicates the changes to the law or regulation, section, or paragraph as "Added 2013" or "Amended 2013." Unless otherwise noted, the effective date of the regulations added or amended in 2013 is January 1, 2014.

Law or Regulation	L&R Committee Item No.	Section	Action	Page
Introduction	BOD Item 120-4	L. Classification for Agenda Items	Amended	4
		M. Developing Items	Deleted and renumbered remaining items	5 to 7
Uniform Weights and Measures Law	221-1	Section 1. Definitions 1.14. Calibration	Amended	22
		1.15. Metrological Traceability	Amended	22
		1.16. Measurement Uncertainty	Amended	22
		1.19. Standard, Reference Measurement	Amended	23
		1.20. Standard, Reference Measurement	Amended	23
		1.21. Standard, Working Measurement	Added	23
		1.22. Metrological Traceability to a Measurement Unit	Added	23
Uniform Regulation for the Method of Sale of Commodities	232-3	2.33. Oil	Amended	133
		2.33.1.5. Tank Trucks or Rail Cars	Amended	134
		2.33.1.6. Documentation	Added	134
		Reference to enforceable date effective July 1, 2013	Deleted	
	232-5	2.34. Retail Sales of Electricity Sold as a Vehicle Fuel	Added	134 to 136
Uniform Engine Fuels and Automotive Lubricants Regulation	237-5	3.13. Oil	Amended	186
		3.13.1.5. Tank Trucks or Rail Cars	Amended	187
		3.13.1.6. Documentation	Added	187
		Reference to enforcement date effective July 1, 2013	Deleted	

2013 Editorial Changes

Law or Regulation	Section	Action	Page
Introduction	Note 1	Updated references	7
II. Uniformity of Laws and Regulations	C. Summary of State Laws and Regulations in Weights and Measures (as of August 1, 2013)	Updated tables and tallies.	10 to 13
Uniform Weights and Measures Law	Note 1	Updated references	23
Uniform Weighmaster Law	Note 1	Updated reference	39
Uniform Packaging and Labeling Regulation	Note 1	Updated reference	61
B. Uniform Regulation for the Method of Sale of Commodities	Note 2	Updated reference	109
	2.10.3. Quantity. Table 1. Softwood Lumber Sizes	Added additional lumber sizes	121
	2.33.1.4. Engine Service Category	Replaced "met" with "displayed"	134
B. Uniform Weighmaster Law	Note 1	Updated references	39
Uniform Engine Fuels and Automotive Lubricants Regulation	3.13.1.4. Engine Service Category	Replaced "met" with "displayed"	187

I. Introduction

A. Source

The Uniform Laws and Regulations[1] in this handbook comprise all of those adopted by the National Conference on Weights and Measures, Inc. (NCWM). The NCWM is supported by the National Institute of Standards and Technology (NIST), which provides its Executive Secretary and publishes its documents. NIST also develops technical publications for use by weights and measures agencies; these publications may subsequently be endorsed or adopted by the NCWM.

The NCWM Committee on Laws and Regulations (the Committee), acting at the request of NCWM or upon its own initiative, prepares with the technical assistance of the National Institute of Standards and Technology (NIST), proposed amendments or additions to the material adopted by NCWM (see Paragraph C). Such revisions, amendments, or additions are then presented to NCWM as a whole where they are discussed by weights and measures officials and representatives of interested manufacturers, industries, consumer groups, and others. Eventually the proposals of the Committee, which may have been amended from those originally presented, are voted upon by the weights and measures officials, following the voting procedures in the NCWM Bylaws. A national consensus is required on all items adopted by the NCWM. A Uniform Law or Regulation is adopted when a majority of the states' representatives, and other voting delegates favoring such adoption, vote for approval.

All of the Uniform Laws and Regulations given herein are recommended by NCWM for adoption by states when reviewing or amending their official laws and regulations in the areas covered. A similar recommendation is made with regard to the local jurisdictions within a state in the absence of the promulgation of such laws and regulations at the state level.

B. Purpose

The purpose of these Uniform Laws and Regulations is to achieve, to the maximum extent possible, uniformity in weights and measures laws and regulations among the various states and local jurisdictions in order to facilitate trade between the states, permit fair competition among businesses, and provide uniform and sufficient protection to all consumers in commercial weights and measures practices.

C. Amendments

The Committee on Laws and Regulations of NCWM serves as a mechanism for consideration of amendments or additions to the Uniform Laws and Regulations.

D. Submission of Agenda Items – Preamble

NCWM Bylaws require that its officers and committees observe the principles of due process for the protection of the rights and interests of affected parties. Specifically, it requires that the committees and officers: (a) give reasonable advance notice of contemplated studies, items to be considered for action, and tentative or definite recommendations for conference vote, and (b) provide that all interested parties have an opportunity to be heard.

[1] *When referring to the Uniform Laws and Regulations in Handbook 130, Laws and Regulations will be capitalized. When referring to general federal or state laws and regulations, no capitalization will be used.*

E. Submission Process

Anyone introducing an item to the Committee must initially use the regional weights and measures associations to consider its merits. Using the regional associations ensures discussion and evaluation of items at the grassroots level by involving the regional members in the development, evaluation, and justification of proposals. The regions include the Central, Northeastern, Southern, and Western Weights and Measures Associations. For information on the regional associations, visit **www ncwm net**.

To submit a proposal to a regional association, obtain *Form 15: Proposal to Amend Handbooks* at **www.ncwm.net** or by contacting NCWM via email at info@ncwm.net. Complete the form and submit it electronically in Microsoft Word format to NCWM at info@ncwm net and copy the Executive Secretary at owm@nist.gov. An example of the Form 15 template is provided at the end of this section. Instructions for completing the form are included with the electronic version of this template. To ensure that your proposal is included on the regional meeting agenda, submit at least two weeks in advance of the fall regional meeting. Regional meeting schedules are available on the NCWM website.

F. Procedures

The NCWM Committee will consider items according to the following procedures:

1. NCWM Committees receive new items from regional associations, National Type Evaluation Technical Committees (Sectors), task groups, and subcommittees and as defined in Sections H and I. All items to be considered by the Committee for action at the upcoming Interim Meeting must be submitted electronically in Microsoft Word format to NCWM by November 1.

2. NCWM will ensure that all committee members and technical advisors receive complete copies of all new items for consideration at the upcoming NCWM Interim Meeting.

G. Criteria for Inclusion on the NCWM Committee's Agenda

1. Any item approved by at least one regional association and received by the November 1 deadline will be automatically placed on the Committee's Interim Meeting agenda.

2. Items that have not been approved by a regional association, but which are received by November 1, will be evaluated by the Committee using the criteria in Section H, Exceptions to Policy, and Section I, Committee Agenda.

3. Any proposal received after the November 1 deadline, but prior to the Interim Meeting, will be evaluated by the Committee according to Section H, Exceptions to Policy and Section I, Committee Agenda. Only those items determined to be a national "priority" will be included on its agenda.

4. Proposals must be in writing and must include:

   a. a concise statement of the item or problem outlining the purpose and national need for its consideration. An electronic copy of the background material and proposed amendment(s) should be submitted in a Microsoft Word format on a CD ROM, DVD, or by electronic mail sent to info@ncwm net;

   b. background material, including test data, analysis of test data, or other appropriately researched and documented material for the Committee to evaluate when deciding its position or future activity on the proposal;

   c. proposed solutions to problems stated in specific language and in amendment form as changes to Conference documents; and

d. if a proposal involves a new area of weights and measures activity; practical, realistic, and specific recommendations for laws or regulations to be adopted and test methods to be utilized to provide for proper enforcement.

When proposals are to modify or add requirements to existing publications, such as Handbook 130, *Uniform Laws and Regulations in the Areas of Legal Metrology and Engine Fuel Quality,* or Handbook 133, *Checking the Net Content of Packaged Goods,* the proposal should:

1) identify the pertinent portion, section, and paragraph of the existing publication that would be changed (e.g., Uniform Method of Sale of Commodities Regulation, Section 8.2, paragraph (b));

2) provide evidence of consistency with other NCWM publications such as with other uniform laws and regulations;

3) provide evidence of consistency with federal laws and regulations (e.g., U.S. Department of Agriculture [USDA] or Federal Trade Commission [FTC] regulations); and

4) relay the positions of businesses, industries, or trade associations affected by the proposal including supporting and opposing points of view.

H. Exceptions to Policy for Submission of Items to the NCWM Committee Agenda; Submission of "Priority" Items

The Committee will use the following criteria to evaluate items that have not been approved by a regional association, but have been received by the November 1 deadline. If an item is received after the November 1 deadline, it will be included on the agenda if the Committee determines that it is a national "priority."

Criteria for Inclusion on the Committee's Agenda When No Regional Association Has Approved the Item.

1. Items must have significant legal impact on weights and measures laws and/or regulations involving:

   a. court cases/attorney general opinions; or

   b. preemption by federal statute or regulation; or

   c. conflicts with international standards; or

   d. items which could affect health and safety.

2. The Committee may contact parties that are potentially affected by an item (e.g., trade associations, industry, and consumer groups) for comments. The Committee may consider these comments and any other information in determining if the item should be included on its agenda.

3. When the Committee determines that it should consider an item as a "priority" (using the criteria in 1.), the item will be handled in the following manner:

   a. A "priority" item received prior to the Interim Meeting may be added to the Interim Meeting agenda by a majority vote of the Committee.

   b. A "priority" item received after the Interim Meeting may be added to the Committee's Annual Meeting agenda as:

      1) a discussion item by a majority vote of the Committee; or

2) as a voting item by a majority vote of the Committee and the NCWM Board of Directors.

I. Committee Agenda

1. The Committee will review items that have been submitted and selected by a majority vote to be included on its agenda. The Committee will only include those items that have been:

   a. approved by at least one of the regional associations; or

   b. forwarded by other committees, subcommittees, NTETC Sectors, task forces, or work groups, or those items that meet the criteria in Section H, Exceptions to Policy.

2. The Committee will publish an agenda (NCWM Publication 15) that identifies the items to be discussed during the Interim Meeting. This agenda will be distributed to members approximately 30 days prior to the meeting. The agenda will be provided upon request to all other interested parties.

(Amended 1998)

J. Interim Meeting

1. The Committee shall hold public hearings at the Interim Meeting for the purpose of discussing and taking comments on all agenda items.

2. Upon request, the Committee will provide the opportunity for presentations by government officials, industry representatives, consumer groups, or other interested parties during the Interim Meeting. Requests to make presentations must be received by the Committee Chairman or Technical Advisor at least two weeks prior to the start of the meetings.

K. Interim Meeting Report

1. Items under consideration by the Committee, and about which the Committee offers comments or recommendations to NCWM to act upon during the Annual Meeting, will be included in the Committee's Interim Report published in the Annual Meeting Program and Committee Reports (NCWM Publication 16).

2. The Annual Meeting Program and Committee Reports will be prepared and distributed to Conference members approximately three months prior to the NCWM Annual Meeting.

L. Classifications for Agenda Items

At the Interim Meeting, the Committee can classify proposals in one of three ways as:

1. "Voting" – These are items the Committee believes are fully developed and ready for final consideration of the voting membership. Each item has either received majority support from the Committee or the Committee has reached agreement that it is ready for voting status to let NCWM membership decide. The Committee has the ability to remove items from the voting agenda at the Annual Meeting by changing the status prior to a vote of the NCWM membership. The Committee may amend voting items during the course of the Annual Meeting based on additional information received following the Interim Meeting and testimony received at the Annual Meeting. These items may also be amended by the voting membership during the voting session of the Annual Meeting following the procedures outlined in the NCWM Bylaws; or

2. "Informational" – These items are deemed by the Committee to have merit. They typically contain a proposal to address the issue at hand and a meaningful background discussion for the proposal. However, the Committee wants to allow more time for review by stakeholders and possibly further development to

address concerns. The Committee has taken the responsibility for any additional development of Informational items. For particularly difficult items, the Committee may assign the item to an existing Subcommittee under its charge or request that the NCWM Chair appoint a special task group that reports to the Committee. At the Annual Meeting, the Committee may change the status of the items, but not to Voting status because the item has not been published as such in advance of the meeting; or

3. "Developing" – These items are deemed by the Committee to have merit, but are found to be lacking enough information for full consideration. Typically the item will have a good explanation of the issue, but a clear proposal has yet to be developed. By assigning Developing status, the Committee has sent the item back to the source or assigned it to some other entity outside the scope of the Committee with the responsibility of further development. The Committee Report will provide the source with clear indication of what is necessary to move the item forward for full consideration. The item will be carried in the Committee agenda in bulletin board fashion with contact information for the person or organization that is responsible for the development. Since the Committee is not required to receive testimony on developing items, this status should be carefully implemented so as not to weaken the standards development process; or

4. "Withdrawn" – These are items that the Committee has found to be without merit. The Committee's determination to withdraw should not be based on the Committee's opinion alone, but on the input received from stakeholders. The Committee's report will contain an explanation for the withdrawal of the item. Once an item appears in NCWM Publication 16 as Withdrawn, the status of that item may not be amended. The item may be reintroduced through the regional associations for consideration as a new item.

(Amended 2013)

M. Comments on Interim Reports

1. Weights and measures officials, industry representatives, and all others are encouraged to submit written comments on items in the Committee's Interim Report.

2. All comments on the Interim Meeting Report must be submitted to the Committee with a copy to the Executive Secretary no later than one month preceding the opening of the Annual Meeting.

N. Annual Meeting

1. The Committee will hold a public hearing at the Annual Meeting to discuss items on its agenda.

2. Those who want to speak on an item during the public hearing should request time from the Committee Chairman. The Committee Chairman may impose time limits on presentations, the discussion of a question, or the discussion of a proposed amendment.

O. Final Committee Reports and Conference Action

1. Following the public hearings, the Committee will prepare its final report for action by the voting membership of the Conference. Copies of the final report will be provided to the membership prior to the voting session for that report.

2. The Chairman of the Committee will present the final report of the Committee to the Conference body. A vote will be taken on items, proposals, or sections in the report as circumstances require. The Conference will vote on the entire final report as presented in accordance with established Conference voting procedures. Parliamentary procedures according to Robert's Rules of Order, as amended by NCWM Bylaws, must be adhered to in the presentation of, and any action on, a Standing Committee report.

   (Amended 1998)

P. Revisions to the Handbook

NIST may not publish a new edition if it determines that it is reasonable to forego an annual publication (e.g., amendments were minor or editorial in nature) to save printing, mailing, and other costs. If this occurs, NIST will issue a notice that the current edition is still valid and will explain its action. (**Note:** Section numbering may be changed from one edition of the handbook to another to accommodate additions or deletions.)
(Amended 2008)

Q. Annotation

Beginning in 1971, amendments or additions to sections in the Uniform Laws and Regulations are delineated at the end of each section (e.g., "amended 1982") as a service to those states that are planning to update their own laws or regulations. The references to each revision and the year will enable legislators and rule makers to study the actual wording and rationale for changes (appearing in the Annual Report of NCWM for that year) and subsequently adopt changes in their own laws and regulations, modeling them after the Uniform Laws and Regulations.

R. Effective Enforcement Dates of Regulations

Unless otherwise specified, the new or amended regulations listed in this section are intended to become effective and subject to enforcement on January 1 of the year following adoption by NCWM.

1. Uniform Packaging and Labeling Regulation
2. Uniform Regulation for the Method of Sale of Commodities
3. Uniform Unit Pricing Regulation
4. Uniform Regulation for the Voluntary Registration of Servicepersons and Service Agencies for Commercial Weighing and Measuring Devices
5. Uniform Open Dating Regulation
6. Uniform Regulation for National Type Evaluation
7. Uniform Regulation for Engine Fuels, Petroleum Products, and Automotive Lubricants

(Added 1992)

S. Section References

In most references made to specific sections or subsections in this handbook, the word "Section" is used, followed by the section number.

T. The International System of Units

The "International System of Units," "SI," or "SI Units" means the modernized metric system as established in 1960 by the General Conference on Weights and Measures (GIPM). In 1988, Congress amended the Metric Conversion Act of 1975 (see Section 5164 of Public Law 100-418) to declare that it is the policy of the United States to designate the metric system of measurement as the preferred measurement system for U.S. trade and commerce, and it further defined "the metric system of measurement" to be the International System as established by the GIPM and as interpreted or modified for the United States by the Secretary of Commerce. [See Metric Conversion Law 15 U.S.C. 205, NIST Special Publication 330 – The International System of Units (SI); NIST Special Publication 814 – Metric System of Measurement; and, Interpretation of the International System of Units for the United States in Federal Register of May 16, 2008, ("Federal Register" Vol. 73, No. 96) or subsequent revisions]. In

1992, Congress amended the Federal Fair Packaging and Labeling Act to require certain consumer commodities to include the appropriate SI units along with the customary inch-pound units in their quantity statements.

(Added 1993) (Revised 2008)

U. "Mass" and "Weight." [*NOTE 1*, page 7]

The mass of an object is a measure of the object's inertial property, or the amount of matter it contains. The weight of an object is a measure of the force exerted on the object by gravity, or the force needed to support it. The pull of gravity on the earth gives an object a downward acceleration of about 9.8 m/s^2. In trade and commerce and everyday use, the term "weight" is often used as a synonym for "mass." The "net mass" or "net weight" declared on a label indicates that the package contains a specific amount of commodity exclusive of wrapping materials. The use of the term "mass" is predominant throughout the world, and is becoming increasingly common in the United States.

(Added 1993)

V. Use of the Terms "Mass" and "Weight." [*NOTE 1*, page 7]

When used in this handbook, the term "weight" means "mass." The term "weight" appears when inch-pound units are cited, or when both inch-pound and SI units are included in a requirement. The terms "mass" or "masses" are used when only SI units are cited in a requirement. The following note appears where the term "weight" is first used in a law or regulation.

NOTE 1: *When used in this law (or regulation), the term "weight" means "mass." (See paragraphs U. "Mass" and Weight and V. Use of the Terms "Mass" and "Weight" in Section I. Introduction of NIST Handbook 130 for an explanation of these terms.)*

(Added 1993)

National Conference on Weights and Measures / National Type Evaluation Program

Form 15: Proposal to Amend Handbooks

General Information (See Instructions)					
1. Date:	**2. Regional Association(s):** __ Central (CWMA)__ Northeastern (NEWMA) __ Southern (SWMA) __ Western (WWMA)				**3. Standing Committee:** __ L&R __ S&T __ PDC
4. Submitter Name:					
5. Street Address:					
6. City:			**7. State:**	**8. Zip Code:**	**9. Country:**
10. Phone Number:	**11. Fax Number:**		**12. E-mail Address:**		
Proposal Information (See Instructions)					
13. Purpose:					
14. Handbook to be Amended: __ *NIST Handbook 44* __ *NIST Handbook 130* __ *NIST Handbook 133* Section: Paragraph:					
15. Proposal:					
16. Justification:					
17. Other Contacts:					
18. Other Reasons For:					
19. Other Reasons Against:					
20. Evidence:					
21. Additional Considerations:					
22. Suggested Action: __ Recommend NCWM Adoption __ Developing Item __ Informational Item __ Other (Please Describe):					
23. List of Attachments:					
For Regional Use Only					
Comments:					

Submit Form Via Email to don.onwiler@ncwm.net: Revised: February 2011

1135 M Street, Suite 110 / Lincoln, Nebraska 68508

P. 402.434.4880 **F.** 402.434.4878 **E.** info@ncwm.net **W.** www.ncwm.net

II. Uniformity of Laws and Regulations

A. National Conference Goal

The goal of the National Conference on Weights and Measures (NCWM) with respect to these Uniform Laws and Regulations is to achieve their acceptance in all states and local jurisdictions that have authority over such matters. The Conference stands ready to assist any jurisdiction in any way possible in securing adoption.

B. Status of Promulgation

The following pages list, by state, information regarding the adoption of the Uniform Laws and Regulations. The tabulated data indicates if the state has adopted the Uniform Law or Regulation by reference, including subsequent amendments (thereby operating under the most recent version of the recommended regulation in this handbook), or if the state has used some version of the NCWM recommended law or regulation as guidance in developing a similar law or regulation.

The information is verified with each state annually; the entries represent the status of the state adoption at the time of the survey.

Unless a state adopts the recommended regulations and subsequent amendments and revisions, there may be variation in the actual degree of adoption. Adoption, implementation, and clarification may be determined by comparing a state law with the Uniform Law, section-by-section, or by contacting the state.

(Amended 1997 and 1998)

C. Summary of State Laws and Regulations in Weights and Measures (as of August 1, 2013)

This is an overview of the status of adoption of NCWM standards by the states. In earlier editions of Handbook 130, state laws and regulations were compared to the NCWM standard from the prior year. This did not indicate whether the standard as printed in the current edition had been adopted by any given state. The table lists those states that adopt NCWM-recommended updates automatically ("YES"); see Sections 4 through 10 and paragraph 12(m) of the Uniform Weights and Measures Law. This means the state's regulations are current with those printed in this edition of the handbook. If a state has adopted an NCWM recommendation in whole or in part from a particular year, but updates are not incorporated automatically, a lower case "yes" is shown. For additional information on the status of adoption, please contact the appropriate state officials.

State	Laws			Regulations									
	Weights and Measures Law	Weighmaster Law or Regulation	Uniform Engine Fuel Law	Packaging and Labeling	Method of Sale	Price Verification	Unit Pricing	Registration of Service Agencies	Open Dating	Type Evaluation	Uniform Engine Fuel Regulation	Handbook 44	Handbook 133
Alabama	yes	yes	yes	yes	yes	YES	NO	yes	NO	yes	yes*	YES	YES
Alaska	yes	NO	NO	YES	YES	yes*	NO	yes	NO	yes	NO	YES	yes
Arizona	yes	yes	yes*	yes	yes	yes	NO	yes	no	yes	yes*	YES	YES
Arkansas	YES	NO	YES	YES	YES	YES	YES	YES	YES	YES	YES	YES	YES
California	yes	yes*	yes*	YES	yes*	yes*	yes*	yes*	NO	yes	yes*	YES	YES
Colorado	yes	yes	yes*	yes	yes	YES	NO	yes	NO	YES	yes*	YES	YES
Connecticut	yes	yes	yes*	YES	YES	YES	yes*	yes*	YES	yes	yes	YES	YES
Delaware	yes	yes	yes*	yes	yes	yes	no	yes*	no	yes	yes*	YES	YES
District of Columbia	yes	yes	NO	yes	yes	no	NO	NO	yes*	no	NO	yes	no
Florida	yes	NO	yes*	yes	yes	yes	yes*	yes	yes*	no	yes*	yes	yes
Georgia	yes	yes	yes*	yes	yes	YES	NO	yes	yes*	yes	yes*	YES	YES
Hawaii	yes	yes	yes*	yes	yes	YES	yes	yes	NO	yes	yes*	yes	yes
Idaho	yes	yes	yes*	yes	yes	YES	no	yes	NO	yes	yes*	YES	yes
Illinois	yes	NO	yes*	YES	YES	NO	NO	yes	NO	yes	yes	YES	YES
Indiana	yes	yes*	yes*	yes	yes	NO	NO	NO	NO	yes	yes*	yes	yes
Iowa	yes	yes*	yes*	yes	yes	YES	NO	yes*	NO	yes	yes*	yes	yes
Kansas	yes	NO	yes	yes	yes	YES	NO	yes	NO	yes	yes	yes	yes

Key:
- YES — Adopted and updated on an annual basis.
- yes — Law or regulation in force, NCWM standard used as basis of adoption, but from an earlier year.
- yes* — Law or regulations in force, but not based on NCWM standard.
- NO — No law or regulation.
- no — No law or regulation, but NCWM standard is used as a guideline.

State	Laws			Regulations									
	Weights and Measures Law	Weighmaster Law or Regulation	Uniform Engine Fuel Law	Packaging and Labeling	Method of Sale	Price Verification	Unit Pricing	Registration of Service Agencies	Open Dating	Type Evaluation	Uniform Engine Fuel Regulation	Handbook 44	Handbook 133
Kentucky	yes	NO	yes*	yes	yes	YES	NO	Yes*	NO	yes	yes*	YES	yes
Louisiana	yes*	yes*	yes*	yes*	NO	YES	no	yes*	NO	yes*	yes*	YES	no
Maine	yes	yes	yes*	YES	YES	YES	NO	yes	NO	yes	YES	YES	no
Maryland	yes	NO	yes*	YES	yes	YES	yes*	yes*	yes*	yes	yes*	YES	YES
Massachusetts	yes*	yes*	yes*	yes	yes*	YES	yes*	NO	NO	yes	yes*	YES	no
Michigan	yes	yes	yes	yes	yes	NO	NO	yes	yes	yes	yes*	yes	yes
Minnesota	yes	NO	yes*	yes*	yes*	NO	NO	yes*	yes*	yes	yes*	yes	yes
Mississippi	yes	yes	yes*	yes	yes	YES	yes	yes	NO	yes	yes*	YES	YES
Missouri	yes	NO	yes	YES	YES	YES	no	yes	NO	YES	yes	YES	YES
Montana	yes	NO	yes	yes	yes	NO	yes	yes	NO	yes	yes	yes	yes
Nebraska	yes	NO	NO	yes	yes	yes*	NO	yes	NO	yes*	NO	yes	yes
Nevada	yes	yes	yes*	YES	YES	YES	YES	YES	YES	YES	yes*	YES	YES
New Hampshire	yes	yes*	NO	YES	YES	YES	YES	yes*	yes*	no	no	YES	YES
New Jersey	yes	yes	yes*	yes	yes	yes*	yes*	yes*	NO	yes*	NO	YES	NO
New Mexico	yes	yes	yes*	yes	yes	YES	NO	yes	yes*	NO	yes*	YES	YES
New York	yes	yes	yes*	yes	yes	YES	yes*	NO	NO	yes	yes*	YES	YES
North Carolina	yes	yes*	yes*	YES	YES	YES	NO	yes	NO	yes	yes*	YES	YES

Key:
- YES — Adopted and updated on an annual basis.
- yes — Law or regulation in force, NCWM standard used as basis of adoption, but from an earlier year.
- yes* — Law or regulations in force, but not based on NCWM standard.
- NO — No law or regulation.
- no — No law or regulation, but NCWM standard is used as a guideline.

	Laws			Regulations									
State	Weights and Measures Law	Weighmaster Law or Regulation	Uniform Engine Fuel Law	Packaging and Labeling	Method of Sale	Price Verification	Unit Pricing	Registration of Service Agencies	Open Dating	Type Evaluation	Uniform Engine Fuel Regulation	Handbook 44	Handbook 133
North Dakota	YES	NO	NO	NO	yes*	NO	NO	yes*	NO	NO	yes*	yes*	NO
Ohio	yes	NO	NO	yes	yes	YES	NO	yes	NO	yes	NO	YES	YES
Oklahoma	YES	NO	yes*	YES	YES	yes*	NO	yes*	YES	YES	yes*	YES	YES
Oregon	yes	NO	yes*	yes	yes	yes	yes*	NO	yes*	yes	yes	yes	yes
Pennsylvania	yes	yes	NO	yes	yes	yes	NO	yes	NO	YES	NO	YES	YES
Puerto Rico	yes	yes	yes*	yes	yes*	yes*	yes*	yes	yes*	yes	yes*	YES	YES
Rhode Island	no	no	yes*	yes*	YES	no	yes*	NO	yes*	no	no	YES	No
South Carolina	yes	yes*	yes*	YES	YES	yes	NO	YES	NO	YES	yes*	YES	YES
South Dakota	yes	NO	yes	yes	yes	yes	NO	yes	yes	yes	yes	yes	yes
Tennessee	yes	yes	yes	YES	YES	YES	NO	yes	NO	YES	yes	YES	YES
Texas	yes	yes*	yes*	YES	YES	yes*	NO	yes	NO	NO	yes*	YES	YES
Utah	yes	NO	yes*	YES	YES	YES	NO	yes	NO	YES	yes	YES	YES
Vermont	yes*	yes	yes*	YES	YES	YES	yes*	yes	NO	no	NO	YES	No
Virginia	yes*	yes*	YES	YES	YES	YES	YES	yes*	NO	yes*	YES	YES	YES
Virgin Islands	yes	NO	yes*	yes	NO	NO	yes	NO	yes	NO	yes	YES	No
Washington	yes	yes	yes	YES	YES	YES	NO	yes	yes	yes	yes	YES	YES
West Virginia	YES	NO	YES	YES	YES	YES	YES	YES	YES	YES	YES	YES	YES

Key:
- YES Adopted and updated on an annual basis.
- yes Law or regulation in force, NCWM standard used as basis of adoption, but from an earlier year.
- yes* Law or regulations in force, but not based on NCWM standard.
- NO No law or regulation.
- no No law or regulation, but NCWM standard is used as a guideline.

State	Laws			Regulations									
	Weights and Measures Law	Weighmaster Law or Regulation	Uniform Engine Fuel Law	Packaging and Labeling	Method of Sale	Price Verification	Unit Pricing	Registration of Service Agencies	Open Dating	Type Evaluation	Uniform Engine Fuel Regulation	Handbook 44	Handbook 133
Wisconsin	yes*	NO	yes*	yes	YES	YES	NO	yes*	NO	YES	yes*	YES	YES
Wyoming	yes	NO	yes*	yes*	no	YES	no	yes	no	YES	yes*	YES	YES
Totals: YES	4	0	3	20	20	30	5	4	5	12	4	40	30
yes	43	21	8	28	25	7	4	28	4	28	11	12	14
yes*	5	10	35	4	5	7	11	14	10	4	29	1	0
NO	0	21	7	1	2	7	28	7	31	4	7	0	2
no	1	1	0	0	1	2	5	0	3	5	2	0	7

Key:
- YES — Adopted and updated on an annual basis.
- yes — Law or regulation in force, NCWM standard used as basis of adoption, but from an earlier year.
- yes* — Law or regulations in force, but not based on NCWM standard.
- NO — No law or regulation.
- no — No law or regulation, but NCWM standard is used as a guideline.

THIS PAGE INTENTIONALLY LEFT BLANK

III. Uniform Laws

Page

THIS PAGE INTENTIONALLY LEFT BLANK

A. Uniform Weights and Measures Law

as adopted by
The National Conference on Weights and Measures*

1. Background

Recognition of the need for uniformity in weights and measures laws and regulations among the states was first noted at the second Annual Meeting of the National Conference on Weights and Measures (NCWM) in April 1906. In the following year, basic outlines of a "Model State Weights and Measures Law" were developed. The first "Model Law," as such, was formally adopted by the Conference in 1911.

Through the years, almost without exception, each state has relied upon the NCWM Weights and Measures Law when the state first enacted comprehensive weights and measures legislation. This has led to a greater degree of uniformity in the basic weights and measures requirements throughout the country.

The original Law was regularly amended to provide for new developments in commercial practices and technology. This resulted in a lengthy and cumbersome document and the need for a simplification of the basic weights and measures provisions. The 1971 NCWM adopted a thoroughly revised, simplified, modernized version of the "Model State Weights and Measures Law." This Law now can serve as a framework for all the many concerns in weights and measures administration and enforcement.

The title of the Law was changed by the 1983 NCWM. Amendments or revisions to the Law since 1971 are noted at the end of each section.

Sections 4 through 10 of the Uniform Weights and Measures Law adopt NIST Handbook 44 and the Uniform Regulations in NIST Handbook 130 by citation. In addition, these sections adopt supplements to and revisions of Handbook 44 and the Uniform Regulations "except insofar as modified or rejected by regulation." Some state laws may not permit enacting a statute that provides for automatic adoption of future supplements to or revisions of a Uniform Regulation covered by that statute. If this should be the case in a given state, two alternatives are available:

(a) Sections 4 through 10 may be enacted without the phrase ". . . and supplements thereto or revisions thereof . . ."; or

(b) Sections 4 through 10 may be enacted by replacing ". . . except insofar as modified or rejected by regulation . . ." with the phrase ". . . as adopted, or amended and adopted, by rule of the director."

Either alternative requires action on the part of the Director to adopt a current version of Handbook 44 and each Uniform Regulation each time a supplement or revision is made by the NCWM.

2. Status of Promulgation

See the table beginning on page 10, Section II. Uniformity of Laws and Regulations of Handbook 130 for the status of adoption of the Uniform Weights and Measures Law.

**The National Conference on Weights and Measures (NCWM) is supported by National Institute of Standards and Technology (NIST) in partial implementation of its statutory responsibility for "cooperation with the states in securing uniformity in weights and measures laws and methods of inspection."*

THIS PAGE INTENTIONALLY LEFT BLANK

Uniform Weights and Measures Law

Table of Contents

Uniform Weights and Measures Law

Section 1. Definitions

When used in this Act:

1.1. Weight(s) and (or) Measure(s). – The term "weight(s) and (or) measure(s)" means all weights and measures of every kind, instruments and devices for weighing and measuring, and any appliance and accessories associated with any or all such instruments and devices.

1.2. Weight. – The term "weight" as used in connection with any commodity or service means net weight. When a commodity is sold by drained weight, the term means net drained weight.
(Amended 1974 and 1990)

1.3. Correct. – The term "correct" as used in connection with weights and measures means conformance to all applicable requirements of this Act.

1.4. Director. – The term "director" means the __________ of the Department of __________.

1.5. Person. – The term "person" means both plural and the singular, as the case demands, and includes individuals, partnerships, corporations, companies, societies, and associations.

1.6. Sale from Bulk. – The term "sale from bulk" means the sale of commodities when the quantity is determined at the time of sale.

1.7. Package. – Except as modified by Section 1. Application of the Uniform Packaging and Labeling Regulation, the term "package," whether standard package or random package, means any commodity:

(a) enclosed in a container or wrapped in any manner in advance of wholesale or retail sale; or

(b) whose weight or measure has been determined in advance of wholesale or retail sale.

An individual item or lot of any commodity on which there is marked a selling price based on an established price per unit of weight or of measure shall be considered a package (or packages).
(Amended 1991)

1.8. Net "Mass" or Net "Weight." – The term "net mass" or "net weight" means the weight [*NOTE 1*, page 21] of a commodity excluding any materials, substances, or items not considered to be part of the commodity. Materials, substances, or items not considered to be part of the commodity include, but are not limited to, containers, conveyances, bags, wrappers, packaging materials, labels, individual piece coverings, decorative accompaniments, and coupons, except that, depending on the type of service rendered, packaging materials may be considered to be part of the service. For example, the service of shipping includes the weight of packing materials.
(Added 1988) (Amended 1989, 1991, and 1993)

1.9. Random Weight Package. – A package that is one of a lot, shipment, or delivery of packages of the same commodity with no fixed pattern of weights.
(Added 1990)

NOTE 1: *When used in this Law, the term "weight" means "mass." (See paragraphs U. "Mass" and "Weight" and V. Use of the Terms "Mass" and "Weight" in Section I. Introduction of NIST Handbook 130 for an explanation of these terms.)*
(Note added 1993)

1.10. Standard Package. – A package that is one of a lot, shipment, or delivery of packages of the same commodity with identical net contents declarations.

Examples:
1 L bottles or 12 fl oz cans of carbonated soda;
500 g or 5 lb bags of sugar;
100 m or 300 ft packages of rope.

(Added 1991) (Amended 1993)

1.11. Commercial Weighing and Measuring Equipment. – The term "commercial weighing and measuring equipment" means weights and measures and weighing and measuring devices commercially used or employed in establishing the size, quantity, extent, area, or measurement of quantities, things, produce, or articles for distribution or consumption, purchased, offered, or submitted for sale, hire, or award, or in computing any basic charge or payment for services rendered on the basis of weight or measure.

(Added 1995)

1.12. Standard, Field. – A physical standard that meets specifications and tolerances in NIST Handbook 105-series standards (or other suitable and designated standards) and is traceable to the reference or working standards through comparisons, using acceptable laboratory procedures, and used in conjunction with commercial weighing and measuring equipment (1.13. Accreditation).

(Added 2005)

1.13. Accreditation. – A formal recognition by a recognized Accreditation Body that a laboratory is competent to carry out specific tests or calibrations or types of tests or calibrations. **NOTE:** Accreditation does not ensure compliance of standards to appropriate specifications.

(Added 2005)

1.14. Calibration. – An operation that, under specified conditions, in a first step, establishes a relation between the quantity values with measurement uncertainties provided by measurement standards and corresponding indications with associated measurement uncertainties and, in a second step, uses this information to establish a relation for obtaining a measurement result from an indication.

(Added 2005) (Amended 2013)

1.15. Metrological Traceability. – The property of a measurement result whereby the result can be related to a reference through a documented unbroken chain of calibrations, each contributing to the measurement uncertainty.

(Added 2005) (Amended 2013)

1.16. Measurement Uncertainty. – A non-negative parameter characterizing the dispersion of the quantity values being attributed to a measurand, based on the information used.

(Added 2005) (Amended 2013)

1.17. Verification. – The formal evaluation of a standard or device against the specifications and tolerances for determining conformance.

(Added 2005)

1.18. Recognition. – A formal recognition by NIST Office of Weights and Measures that a laboratory has demonstrated the ability to provide traceable measurement results and is competent to carry out specific tests or calibrations or types of tests or calibrations.

(Added 2005)

1.19. Standard, Reference Measurement. – A measurement standard designated for the calibration of other measurement standards for quantities of a given kind in a given organization or at a given location. The term "reference measurement standards" usually means the physical standards of the state that serve as the legal reference from which all other standards for weights and measures within that state are derived.

(Added 2005) (Amended 2013)

1.20. Standard, Working Measurement. – A measurement standard that is used routinely to calibrate or verify measuring instruments or measuring systems. The term "working measurement standards" means the physical standards that are traceable to the reference standards through calibrations or verifications, using acceptable laboratory procedures, and used in the enforcement of weights and measures laws and regulations.

(Added 2005) (Amended 2013)

1.21. Metrological Traceability Chain. – Sequence of measurement standards and calibrations that is used to relate a measurement result to a reference.

(Added 2013)

1.22. Metrological Traceability to a Measurement Unit. – Metrological traceability where the reference is the definition of a measurement unit through its practical realization.

(Added 2013)

Section 2. Systems of Weights and Measures

The International System of Units (SI) and the system of weights and measures in customary use in the United States are jointly recognized, and either one or both of these systems shall be used for all commercial purposes in the state.

The definitions of basic units of weight and measure, the tables of weight and measure, and weights and measures equivalents as published by NIST are recognized and shall govern weighing and measuring equipment and transactions in the state.

(Amended 1993)

NOTE 2: SI or SI Unit. *– means the International System of Units as established in 1960 by the General Conference on Weights and Measures and interpreted or modified for the United States by the Secretary of Commerce. See "Interpretation of the International System of Units for the United States" in "Federal Register" (Volume 73, No. 96, pages 28432 to 28433) for May 16, 2008, and 15 United States Code, Section 205a - 205l "Metric Conversion." See also NIST Special Publication 330, "The International System of Units (SI)," 2008 edition and NIST Special Publication 811, "Guide for the Use of the International System of Units (SI)," 2008 edition that are available at* **www nist.gov/pml/wmd/metric/metric-program.cfm** *or by contacting TheSI@nist.gov.*

(Added 1993)

Section 3. Physical Standards

Weights and measures that are traceable to the U.S. prototype standards supplied by the Federal Government, or approved as being satisfactory by NIST, shall be the state reference and working standards of weights and measures, and shall be maintained in such calibration as prescribed by the NIST as demonstrated through laboratory accreditation or recognition. All field standards may be prescribed by the Director and shall be verified upon their initial receipt and as often thereafter as deemed necessary by the Director.

(Amended 2005)

Section 4. Technical Requirements for Weighing and Measuring Devices [*NOTE 3*, page 24]

The specifications, tolerances, and other technical requirements for commercial, law enforcement, data gathering, and other weighing and measuring devices as adopted by the NCWM, published in the National Institute of Standards and Technology Handbook 44, "Specifications, Tolerances, and Other Technical Requirements for Weighing and Measuring Devices," and supplements thereto or revisions thereof, shall apply to weighing and measuring devices in the state, except insofar as modified or rejected by regulation.

(Amended 1975)

NOTE 3: *Sections 4 through 10 of the Uniform Weights and Measures Law adopt NIST Handbook 44 and Uniform Regulations in NIST Handbook 130 by citation. In addition, these sections adopt supplements to and revisions of NIST Handbook 44 and the Uniform Regulations "except insofar as modified or rejected by regulation." Some state laws may not permit enacting a statute that provides for automatic adoption of future supplements to or revisions of a regulation covered by that statute. If this should be the case in a given state, two alternatives are available:*

(a) Sections 4 through 10 may be enacted without the phrase ". . . and supplements thereto or revisions thereof . . ."; or

(b) Sections 4 through 10 may be enacted by replacing ". . . except insofar as modified or rejected by regulation . . ." with the phrase ". . . as adopted, or amended and adopted, by rule of the director."

Either alternative requires action on the part of the Director to adopt a current version of Handbook 44 and Uniform Laws or Regulations each time a supplement is added or revision is made by the NCWM.

Section 5. Requirements for Packaging and Labeling [*NOTE 3*, page 24]

The Uniform Packaging and Labeling Regulation as adopted by the NCWM and published in the National Institute of Standards and Technology Handbook 130, "Uniform Laws and Regulations," and supplements thereto or revisions thereof, shall apply to packaging and labeling in the state, except insofar as modified or rejected by regulation.

(Added 1983)

Section 6. Requirements for the Method of Sale of Commodities [*NOTE 3*, page 24]

The Uniform Regulation for the Method of Sale of Commodities as adopted by the NCWM and published in National Institute of Standards and Technology Handbook 130, "Uniform Laws and Regulations," and supplements thereto or revisions thereof, shall apply to the method of sale of commodities in the state, except insofar as modified or rejected by regulation.

(Added 1983)

Section 7. Requirements for Unit Pricing [*NOTE 3*, page 24]

The Uniform Unit Pricing Regulation as adopted by the NCWM and published in the National Institute of Standards and Technology Handbook 130, "Uniform Laws and Regulations," and supplements thereto or revisions thereof, shall apply to unit pricing in the state, except insofar as modified or rejected by regulation.

(Added 1983)

Section 8. Requirements for the Registration of Servicepersons and Service Agencies for Commercial Weighing and Measuring Devices [*NOTE 3*, page 24]

The Uniform Regulation for the Voluntary Registration of Servicepersons and Service Agencies for Commercial Weighing and Measuring Devices as adopted by the National NCWM and published in the National Institute of Standards and Technology Handbook 130, "Uniform Laws and Regulations," and supplements thereto or revisions thereof, shall apply to the registration of servicepersons and service agencies in the state, except insofar as modified or rejected by regulation.

(Added 1983)

Section 9. Requirements for Open Dating [*NOTE 3*, page 24]

The Uniform Open Dating Regulation as adopted by the NCWM and published in the National Institute of Standards and Technology Handbook 130, "Uniform Laws and Regulations," and supplements thereto or revisions thereof, shall apply to open dating in the state, except insofar as modified or rejected by regulation.

(Added 1983)

Section 10. Requirements for Type Evaluation [*NOTE 3*, page 24]

The Uniform Regulation for National Type Evaluation as adopted by the NCWM and published in National Institute of Standards and Technology Handbook 130, "Uniform Laws and Regulations," and supplements thereto or revisions thereof, shall apply to type evaluation in the state, except insofar as modified or rejected by regulation.

(Added 1985)

Section 11. State Weights and Measures Division

There shall be a State Division of Weights and Measures located for administrative purposes within the Department of _________ (agency, etc.). The Division is charged with, but not limited to, performing the following functions on behalf of the citizens of the state:

(a) Assuring that weights and measures in commercial services within the state are suitable for their intended use, properly installed, and accurate, and are so maintained by their owner or user.

(b) Preventing unfair or deceptive dealing by weight or measure in any commodity or service advertised, packaged, sold, or purchased within the state.

(c) Making available to all users of physical standards or weighing and measuring equipment the precision calibration and related metrological certification capabilities of the weights and measures facilities of the Division.

(d) Promoting uniformity, to the extent practicable and desirable, between weights and measures requirements of this state and those of other states and federal agencies.

(e) Encouraging desirable economic growth while protecting the consumer through the adoption by rule of weights and measures requirements as necessary to assure equity among buyers and sellers.

(Added 1976)

Section 12. Powers and Duties of the Director

The Director shall:

(a) maintain traceability of the state standards as demonstrated through laboratory accreditation or recognition;

(Amended 2005)

(b) enforce the provisions of this Act;

(c) issue reasonable regulations for the enforcement of this Act, which regulations shall have the force and effect of law;

(d) establish labeling requirements, establish requirements for the presentation of cost per unit information, establish standards of weight, measure, or count, and reasonable standards of fill for any packaged commodity; and establish requirements for open dating information;

(Added 1973)

(e) grant any exemptions from the provisions of this Act or any regulations promulgated pursuant thereto when appropriate to the maintenance of good commercial practices within the state;

(f) conduct investigations to ensure compliance with this Act;

(g) delegate to appropriate personnel any of these responsibilities for the proper administration of this office;

(h) verify the field standards for weights and measures used by any jurisdiction within the state, before being put into service, tested annually or as often thereafter as deemed necessary by the Director based on statistically evaluated data, and approve the same when found to be correct;

(Amended 2005)

(i) have the authority to inspect and test commercial weights and measures kept, offered, or exposed for sale;

(Amended 1995)

(j) inspect and test, to ascertain if they are correct, weights and measures commercially used:

(1) in determining the weight, measure, or count of commodities or things sold, or offered or exposed for sale, on the basis of weight, measure, or count; or

(2) in computing the basic charge or payment for services rendered on the basis of weight, measure, or count.

(k) test all weights and measures used in checking the receipt or disbursement of supplies in every institution, the maintenance of which funds are appropriated by the legislature of the state;

(l) approve for use, and may mark, such commercial weights and measures as are found to be correct, and shall reject and order to be corrected, replaced, or removed such commercial weights and measures as are found to be incorrect. Weights and measures that have been rejected may be seized if not corrected within the time specified or if used or disposed of in a manner not specifically authorized. The Director shall remove from service and may seize the weights and measures found to be incorrect that are not capable of being made correct;

(Amended 1995)

(m) weigh, measure, or inspect packaged commodities kept, offered, or exposed for sale, sold, or in the process of delivery, to determine whether they contain the amounts represented and whether they are kept, offered, or exposed for sale in accordance with this Act or regulations promulgated pursuant thereto. In carrying out the provisions of this section, the Director shall employ recognized sampling procedures, such as are adopted by the NCWM and are published in the National Institute of Standards and Technology Handbook 133, "Checking the Net Contents of Packaged Goods;"

(Amended 1984, 1988, and 2000)

(n) prescribe, by regulation, the appropriate term or unit of weight or measure to be used, whenever the Director determines that an existing practice of declaring the quantity of a commodity or setting charges for a service by weight, measure, numerical count, time, or combination thereof, does not facilitate value comparisons by consumers, or offers an opportunity for consumer confusion;

(Amended 1991)

(o) allow reasonable variations from the stated quantity of contents, which shall include those caused by loss or gain of moisture during the course of good distribution practice or by unavoidable deviations in good manufacturing practice only after the commodity has entered intrastate commerce;

(p) provide for the training of weights and measures personnel, and may establish minimum training and performance requirements, which shall then be met by all weights and measures personnel, whether county, municipal, or state. The Director may adopt the training standards of the National Conference on Weights and Measures' National Training Program and the laboratory metrology standards specified by the NIST accreditation and/or recognition requirements; and

(Added 1991) (Amended 2005)

(q) verify advertised prices, price representations, and point-of-sale systems, as deemed necessary, to determine:

(1) the accuracy of prices and computations and the correct use of the equipment; and

(2) if such system utilizes scanning or coding means in lieu of manual entry, the accuracy of prices printed or recalled from a database. In carrying out the provisions of this section, the Director shall:

i. employ recognized procedures, such as are designated in National Institute of Standards and Technology Handbook 130, *Uniform Laws and Regulations in the Areas of Legal Metrology and Engine Fuel Quality*, "Examination Procedures for Price Verification;"

ii. issue necessary rules and regulations regarding the accuracy of advertised prices and automated systems for retail price charging (referred to as "point-of-sale systems") for the enforcement of this section, which rules shall have the force and effect of law; and

iii. conduct investigations to ensure compliance.

(Added 1995)

Section 13. Special Police Powers

When necessary for the enforcement of this Act or regulations promulgated pursuant thereto, the Director is:

(a) Authorized to enter any commercial premises during normal business hours, except that in the event such premises are not open to the public, he/she shall first present his/her credentials and obtain consent before making entry thereto, unless a search warrant has previously been obtained.

(b) Empowered to issue stop use, hold, and removal orders with respect to any weights and measures commercially used, stop sale, hold, and removal orders with respect to any packaged commodities or bulk commodities kept, offered, or exposed for sale.

(c) Empowered to seize, for use as evidence, without formal warrant, any incorrect or unapproved weight, measure, package, or commodity found to be used, retained, offered, or exposed for sale or sold in violation of the provisions of this Act or regulations promulgated pursuant thereto.

(d) Empowered to stop any commercial vehicle and, after presentation of his credentials, inspect the contents, require the person in charge of that vehicle to produce any documents in his possession concerning the contents, and require him to proceed with the vehicle to some specified place for inspection.

(e) With respect to the enforcement of this Act, the Director is hereby vested with special police powers, and is authorized to arrest, without formal warrant, any violator of this Act.

Section 14. Powers and Duties of Local Officials

Any weights and measures official appointed for a county or city shall have the duties and powers enumerated in this Act, excepting those duties reserved to the state by law or regulation. These powers and duties shall extend to their respective jurisdictions, except that the jurisdiction of a county official shall not extend to any city for which a weights and measures official has been appointed. No requirement set forth by local agencies may be less stringent than or conflict with the requirements of the state.
(Amended 1984)

Section 15. Misrepresentation of Quantity

No person shall:

(a) sell, offer, or expose for sale a quantity less than the quantity represented; nor

(b) take more than the represented quantity when, as buyer, he/she furnishes the weight or measure by means of which the quantity is determined; nor

(c) represent the quantity in any manner calculated or tending to mislead or in any way deceive another person.
(Amended 1975 and 1990)

Section 16. Misrepresentation of Pricing

No person shall misrepresent the price of any commodity or service sold, offered, exposed, or advertised for sale by weight, measure, or count, nor represent the price in any manner calculated or tending to mislead or in any way deceive a person.

Section 17. Method of Sale

Except as otherwise provided by the Director or by firmly established trade custom and practice,

(a) commodities in liquid form shall be sold by liquid measure or by weight; and

(b) commodities not in liquid form shall be sold by weight, by measure, or by count.

The method of sale shall provide accurate and adequate quantity information that permits the buyer to make price and quantity comparisons.
(Amended 1989)

Section 18. Sale from Bulk

All bulk sales in which the buyer and seller are not both present to witness the measurement, all bulk deliveries of heating fuel, and all other bulk sales specified by rule or regulation of the director shall be accompanied by a delivery ticket containing the following information:

(a) the name and address of the buyer and seller;

(b) the date delivered;

(c) the quantity delivered and the quantity upon which the price is based, if this differs from the delivered quantity for example, when temperature compensated sales are made;
(Amended 1991)

(d) the unit price, unless otherwise agreed upon by both buyer and seller;
(Added 1991)

(e) the identity in the most descriptive terms commercially practicable, including any quality representation made in connection with the sale; and

(f) the count of individually wrapped packages, if more than one, in the instance of commodities bought from bulk but delivered in packages.
(Amended 1983 and 1991)

Section 19. Information Required on Packages

Except as otherwise provided in this Act or by regulations promulgated pursuant thereto, any package, whether a random package or a standard package, kept for the purpose of sale, or offered or exposed for sale, shall bear on the outside of the package a definite, plain, and conspicuous declaration of:

(a) the identity of the commodity in the package, unless the commodity is a food, other than meat or poultry, that was repackaged in a retail establishment and the food is displayed to the purchaser under either of the following circumstances:

(1) its interstate labeling is clearly in view or with a counter card, sign or other appropriate device bearing prominently and conspicuously the common or usual name of the food; or

(2) the common or usual name of the food is clearly revealed by its appearance.
(Amended 2001)

(b) the quantity of contents in terms of weight, measure, or count; and,

(c) the name and place of business of the manufacturer, packer, or distributor, in the case of any package kept, offered, or exposed for sale, or sold in any place other than on the premises where packed.
(Amended 1991)

Section 20. Declarations of Unit Price on Random Weight Packages

In addition to the declarations required by Section 19. Information Required on Packages of this Act, any package being one of a lot containing random weights of the same commodity, at the time it is offered or exposed for sale at retail, shall bear on the outside of the package a plain and conspicuous declaration of the price per kilogram or pound and the total selling price of the package.
(Amended 1986)

Section 21. Advertising Packages for Sale

Whenever a packaged commodity is advertised in any manner with the retail price stated, there shall be closely and conspicuously associated with the retail price a declaration of quantity as is required by law or regulation to appear on the package.
(Amended 1993)

Section 22. Prohibited Acts

No person shall:

(a) use or have in possession for use in commerce any incorrect weight or measure;

(b) sell or offer for sale for use in commerce any incorrect weight or measure;

(c) remove any tag, seal, or mark from any weight or measure without specific written authorization from the proper authority;

(d) hinder or obstruct any weights and measures official in the performance of his or her duties; or

(e) violate any provisions of this Act or regulations promulgated under it.

Section 23. Civil Penalties

23.1. Assessment of Penalties. – Any person who by himself or herself, by his or her servant or agent, or as the servant or agent of another person, commits any of the acts enumerated in Section 22. Prohibited Acts may be assessed by the __________ a civil penalty of:

(a) not less than $__________ nor more than $__________ for a first violation;

(b) not less than $__________ nor more than $__________ for a second violation within __________ from the date of the first violation; and

(c) not less than $__________ nor more than $__________ for a third violation within __________ from the date of the first violation.

23.2. Administrative Hearing. – Any person subject to a civil penalty shall have a right to request an administrative hearing within __________ days of receipt of the notice of the penalty. The Director or his/her designee shall be authorized to conduct the hearing after giving appropriate notice to the respondent. The decision of the Director shall be subject to appropriate judicial review.

23.3. Collection of Penalties. – If the respondent has exhausted his or her administrative appeals and the civil penalty has been upheld, he or she shall pay the civil penalty within __________ days after the effective date of the final decision. If the respondent fails to pay the penalty, a civil action may be brought by the Director in any court of competent jurisdiction to recover the penalty. Any civil penalty collected under this Act shall be transmitted to __________.

(Added 1989) (Amended 1995)

Section 24. Criminal Penalties

24.1. Misdemeanors. – Any person who commits any of the acts enumerated in Section 22. Prohibited Acts shall be guilty of a Class __________ misdemeanor and upon a first conviction thereof shall be punished by a fine of not less than $__________ nor more than $__________ or by imprisonment for not more than __________ months, or both. Upon a subsequent conviction thereof, he or she shall be punished by a fine of not less than $__________ nor more than $__________ or by imprisonment for up to __________, or both.

24.2. Felonies. – Any person who:

(a) intentionally violates any provisions of this Act or regulations under it;

(b) is convicted under the misdemeanor provisions of Section 24(a) more than three times in a 2-year period; or

(c) uses or has in his or her possession a device which has been altered to facilitate fraud shall be guilty of a Class __________ felony and upon a first offense shall be punished by a fine of not less than $__________, or by imprisonment for not more than __________, or both.

(Added 1989)

Section 25. Restraining Order and Injunction

The Director is authorized to apply to any court of competent jurisdiction for a restraining order, or a temporary or permanent injunction, restraining any person from violating any provision of this Act.

(Retitled 1989)

Section 26. Presumptive Evidence

Whenever there shall exist a weight or measure or weighing or measuring device in or about any place in which or from which buying or selling is commonly carried on, there shall be a rebuttable presumption that such weight or measure or weighing or measuring device is regularly used for the business purposes of that place.

Section 27. Separability Provision

If any provision of this Act is declared unconstitutional, or the applicability thereof to any person or circumstance is held invalid, the constitutionality of the remainder of the Act and the applicability thereof to other persons and circumstances shall not be affected thereby.

Section 28. Repeal of Conflicting Laws

All laws and parts of laws contrary to or inconsistent with the provisions of this Act are repealed except as to offenses committed, liabilities incurred, and claims made there under prior to the effective date of this Act.

Section 29. Regulations to be Unaffected by Repeal of Prior Enabling Statute

The adoption of this Act or any of its provisions shall not affect any regulations promulgated pursuant to the authority of any earlier enabling statute unless inconsistent with this Act or modified or revoked by the Director.

Section 30. Effective Date

This Act shall become effective on __________.

THIS PAGE INTENTIONALLY LEFT BLANK

B. Uniform Weighmaster Law

as adopted by
The National Conference on Weights and Measures*

1. Background

The "Model State Weighmaster Law" was first proposed by the 35th National Conference on Weights and Measures (NCWM) in 1950. It was formally adopted by the NCWM and recommended to the states in 1951.

Over the years, very few changes have been made to the Weighmaster Law until 1965. In that year, the format of the Weighmaster Law was revised to be in full accord with the Weights and Measures Law. The name was changed to "Uniform Weighmaster Law" in 1983. The law was again completely revised and updated in 1989. It was editorially revised in 1990.

It provides a registration, licensing, and enforcement program for "public weighmasters" or third-party measurers in commercial transactions.

2. Status of Promulgation

See the table beginning on page 10, Section II. Uniformity of Laws and Regulations of Handbook 130, *Uniform Laws and Regulations in the Areas of Legal Metrology and Fuel Quality,* for the status of adoption of the Uniform Weighmaster Law.

**The National Conference on Weights and Measures (NCWM) is supported by the National Institute of Standards and Technology (NIST) in partial implementation of its statutory responsibility for "cooperation with the states in securing uniformity in weights and measures laws and methods of inspection."*

THIS PAGE INTENTIONALLY LEFT BLANK

Uniform Weighmaster Law

Table of Contents

Uniform Weighmaster Law

Section 1. Purpose

The purpose of this Act is to ensure accurate measurements by public weighmasters.

Section 2. Scope

This Act:

(a) establishes a registration, licensing, and enforcement program;

(b) provides authority for license fee collection;

(c) empowers the state to promulgate regulations as needed to carry out the provisions of the Act;

(d) provides for optional or voluntary licensing when the employing organization or other organizations require it as part of the condition for employment; and

(e) provides for civil and criminal penalties.

Section 3. Definitions

As used in this Act:

3.1. Public Weighing. – The weighing, measuring, or counting, upon request, of vehicles, property, produce, commodities, or articles other than those that the weigher or his/her employer, if any, is either buying or selling.

3.2. Public Weighmaster. – Any person who performs public weighing as defined in 3.1. Public Weighing.

3.3. Vehicle. – Any device (except railroad freight cars) in, upon, or by which any property, produce, commodity, or article is or may be transported or drawn.

3.4. Director. – The __________ of the Department of __________.

Section 4. Enforcing Officer: Rules and Regulations

The Director is authorized to:

(a) enforce the provisions of this Act;

(b) issue reasonable regulations for the enforcement of this Act that shall have the force and effect of law; and

(c) adopt rules that include, but are not limited to, determining:

(1) the qualifications of the applicant for a license as a public weighmaster;

(2) renewal or refusal of a license;

(3) the period of license validity;

(4) measurement practices that must be followed, including the measurement or recording of tare;

(5) the required information to be submitted with or as part of a certificate; and

(6) the period of recordkeeping.

Section 5. Qualifications for Weighmaster

To receive authorization to act as a public weighmaster, a person must receive a license from the Director. In order to qualify for a license, a person must:

(a) be able to weigh or measure accurately;

(b) be able to generate correct certificates; and

(c) possess other qualifications required by regulations promulgated under the Act.

Section 6. License Application

Using a form provided by the Director, the applicant for a license as a public weighmaster shall furnish evidence that he/she has the qualifications required by Section 5. Qualifications for Weighmaster of this Act and regulations promulgated under the Act.

Section 7. Evaluation of Qualifications of Applicants

The Director will determine the qualifications of the applicant based on:

(a) the information provided on the application; and

(b) supplementary information as determined by the Director.

The Director may also determine the qualifications of the applicant based on the results of an examination of the applicant's knowledge.

Section 8. Issuance and Records of Licenses

The Director will:

(a) grant licenses as public weighmasters to qualified applicants;

(b) keep a record of all applications submitted and of all licenses issued; and

(c) establish the period of validity of licenses issued.

Section 9. License Fees

The Director shall have the authority to set fees for the administration and effective enforcement of the provisions of this Act. Before the issuance of a new license or renewal of a license as a public weighmaster, the applicant must pay a fee of $__________ to the Director.

Section 10. Certificate: Required Entries

(a) The certificate, when properly filled out and signed shall be prima facie evidence of the accuracy of the measurements shown.

(b) The design of and the information to be furnished on a weight certificate shall be prescribed by the Director and will include, but not be limited to, the following:

(1) the name and license number of the public weighmaster;

(2) the kind of commodity weighed, measured, or counted;

(3) the name of the owner, agent, or consignee of the commodity;

(4) the name of the recipient of the commodity, if applicable;

(5) the date the certificate is issued;

(6) the consecutive number of the certificate;

(7) the identification, including the identification number, if any, of the carrier transporting the commodity and the identification number or license number of the vehicle;

(8) other information needed to distinguish or identify the commodity from a like kind;

(9) the number of units of the commodity, if applicable;

(10) the measure of the commodity, if applicable;

(11) the weight [*NOTE 1*, page 39] of the commodity and the vehicle or container (if applicable) broken down as follows:

i. the gross weight of the commodity and the associated vehicle or container;

ii. the tare weight of the unladened vehicle or container; or

iii. both the gross and tare weight and the resultant net weight of the commodity;

(12) signature of the public weighmaster who determined the weight, measure, or count.

NOTE 1: *When used in this Law, the term "weight" means "mass." (See paragraph U. "Mass" and "Weight"in Section I. Introduction, of NIST Handbook 130 for an explanation of these terms.)*
(Note added 1993)

Section 11. Certificate: Execution, Requirements

(a) When filling out a certificate, a public weighmaster shall:

(1) enter the measurement values to clearly show that the measurements were actually determined;

(2) enter only the measurement values personally determined; and

(3) not enter measurement values determined by other persons.

(b) If the certificate provides for entries of gross, tare, or net, the public weighmaster shall:

(1) strike out or otherwise cancel the printed entries for the values not determined; or

(2) enter the scale and date on which the values were determined on the certificate if the values were not determined on the same scale or on the same date shown on the certificate.

Section 12. Measurement Practices and Equipment Used

A public weighmaster shall use measurement practices and equipment:

(a) in accordance with the requirements of the latest edition of NIST Handbook 44, "Specifications, Tolerances, and Other Technical Requirements for Weighing and Measuring Devices"; and

(b) examined, tested, and approved for use by a weights and measures officer of this state.

Section 13. Scale Used: Capacity, Platform Size, One-Draft Weighing

(a) A public weighmaster shall not weigh a vehicle, or combination of vehicles, when part of the vehicle or connected combination, is not resting fully, completely, and as one entire unit on the scale.

(b) When weighing a combination of vehicles that will not rest fully, completely, and as one complete unit on the scale platform:

   (1) the combination shall be disconnected and weighed in single drafts; and

   (2) the weights of the single drafts may be combined in order to issue a single certificate for the combination, provided the certificate indicates that the total represents a combination of single draft weighings.

Section 14. Copies of Certificates

A public weighmaster shall keep and preserve for the period specified in the regulations a legible copy of each certificate issued by him or her. The certificates shall be available for inspection by any weights and measures officer of this state during normal office hours.

Section 15. Reciprocal Acceptance of Certificates

The Director is authorized to recognize and accept certificates issued by licensed public weighmasters of other states that recognize and accept certificates issued by licensed weighmasters of this state.

Section 16. Optional Licensing

The following persons shall be authorized, but are not required, to obtain licenses as public weighmasters:

(a) a law enforcement or weights and measures officer or other qualified employee of a state, city, or county agency or institution when acting within the scope of his/her official duties;

(b) a person weighing property, produce, commodities, or articles:

   (1) that he/she or his/her employer is either buying or selling; or

   (2) in conformity with the requirements of federal statutes or the statutes of this state relative to warehousemen or processors.

Section 17. Prohibited Acts

It is a prohibited act for any person

(a) without a valid license to:

(1) assume the title of public weighmaster or any title of similar import;

(2) perform the duties or acts to be performed by a public weighmaster;

(3) hold himself or herself out as a public weighmaster;

(4) issue any certificate, ticket, memorandum, or statement for which a fee is charged; or

(5) engage in full-time or part-time business of measuring for hire.

(b) to use or operate any device for certification purposes that does not meet, nor in a manner not in accordance with, the requirements of the latest edition of NIST Handbook 44, "Specifications, Tolerances, and Other Technical Requirements for Weighing and Measuring Devices";

(c) to falsify a certificate or to falsely certify any gross, tare, or net weight or measure required by the Act to be on the certificate;

(d) to refuse without cause to weigh or measure any article or thing which it is his/her duty to weigh or measure, or refuse to state in any certificate anything required to be therein;

(e) to hinder or obstruct in any way the Director or his/her authorized agent in the performance of the Director's official duties under this Act;

(f) to violate any provision of this Act or any regulation promulgated under this Act;

(g) to delegate his/her authority to any person not licensed as a public weighmaster;

(h) to request a false certificate or to request a public weighmaster to weigh, measure, or count any vehicle, property, produce, commodity, or article falsely or incorrectly;

(i) to issue a certificate simulating the certificate in the Act; or

(j) to use or have in his/her possession a device which has been altered to facilitate fraud.

Section 18. Suspension and Revocation of License

The Director is authorized to suspend or revoke the license of any public weighmaster:

(a) when, after a hearing held following 10 days' notice to the licensee, he/she is satisfied that the licensee has violated any provision of this Act or any regulation under this Act;

(b) when the licensee has been convicted in any court of competent jurisdiction of violating any provision of this Act or any regulation under this Act; or

(c) when the licensee is convicted of any felony.

Section 19. Civil Penalties

19.1. Assessment of Penalties. – Any person who by himself or herself, by his or her servant or agent, or as the servant or agent of another person commits any of the acts enumerated in Section 22. Validity of Prosecutions may be assessed by the __________ a civil penalty of:

(a) not less than $______ nor more than $______ for a first violation,

(b) not less than $_____ nor more than $_____ for a second violation within _____ from the date of the first violation, and

(c) not less than $______ nor more than $______ for a third violation within ______ from the date of the first violation.

19.2. Administrative Hearing. – Any person subject to a civil penalty shall have a right to request an administrative hearing within __________ days of receipt of the notice of the penalty. The Director or his/her designee shall be authorized to conduct the hearing after giving appropriate notice to the respondent. The decision of the Director shall be subject to appropriate judicial review.

19.3. Collection of Penalties. – If the respondent has exhausted his or her administrative appeals and the civil penalty has been upheld, he or she shall pay the civil penalty within __________ days after the effective date of the final decision. If the respondent fails to pay the penalty, a civil action may be brought by the Director in any court of competent jurisdiction to recover the penalty. Any civil penalty collected under this Act shall be transmitted to __________.

(Added 1989) (Amended 1995)

Section 20. Criminal Penalties

20.1. Misdemeanor. – Any person who by himself/herself, by his/her servant or agent, or as the servant or agent of another person commits any of the acts enumerated in Section 17. Prohibited Acts or violates any other provision of this Act shall be guilty of a Class __________ misdemeanor and upon conviction shall be punished by a fine not less than $__________, nor more than $__________, or by imprisonment for not less than __________ nor more than __________, or both fine and imprisonment.

20.2. Felony. – Any person who by himself/herself, by his/her servant or agent, or as the servant or agent of another person intentionally commits any of the acts enumerated in Section 17. Prohibited Acts or repeatedly violates any other provision of this Act shall be guilty of a Class __________ felony and upon conviction shall be punished by a fine not less than $__________ and/or by imprisonment for not less than __________, nor more than __________, or more than __________.

Section 21. Restraining Order and Injunction

The Director is authorized to apply to any court of competent jurisdiction for a restraining order, or a temporary or permanent injunction, restraining any person from violating any provision of this Act.

Section 22. Validity of Prosecutions

Prosecutions for violation of any provision of this Act are declared to be valid and proper notwithstanding the existence of any other valid general or specific Act of this state dealing with matters that may be the same as or similar to those covered by this Act.

Section 23. Separability Provision

If any provision of this Act is declared unconstitutional, or the applicability thereof to any person or circumstance is held invalid, the constitutionality of the remainder of the Act and the applicability thereof to other persons and circumstances shall not be affected.

Section 24. Repeal of Conflicting Laws

All laws and parts of laws contrary to or inconsistent with the provisions of this Act, and specifically __________, are repealed insofar as they might operate in the future; but as to offenses committed, liabilities incurred, and claims now existing there under, the existing law shall remain in full force and effect.

Section 25. Citation

This Act may be cited as the "Public Weighmaster Act of __________."

Section 26. Effective Date

This Act shall become effective on __________.

THIS PAGE INTENTIONALLY LEFT BLANK

C. Uniform Engine Fuels and Automotive Lubricants Inspection Law

as adopted by
The National Conference on Weights and Measures*

1. Background

In 1984, the National Conference on Weights and Measures (NCWM) adopted a section in the Uniform Regulation for the Method of Sale of Commodities requiring that motor fuel containing alcohol be labeled to disclose to the retail purchaser that the fuel contains alcohol. The delegates deemed this action necessary since motor vehicle manufacturers were qualifying their warranties with respect to some gasoline-alcohol blends, motor fuel users were complaining to weights and measures officials about fuel quality and vehicle performance, and the American Society for Testing and Materials (ASTM) had not yet finalized quality standards for oxygenated (which includes alcohol-containing) fuels. While many argued that weights and measures officials should not cross the line from quantity assurance programs to programs regulating quality, the delegates were persuaded that the issue needed immediate attention.

A Motor Fuels Task Force was appointed in 1984 to develop mechanisms for achieving uniformity in the evaluation and regulation of motor fuels. The Task Force developed the Uniform Motor Fuel Inspection Law and the Uniform Motor Fuel Regulation (see the Uniform Regulations section of this Handbook) to accompany the Law. The recommended Law required registration and certification of motor fuel as meeting ASTM standards. It established a motor fuel quality testing capability by the state. Funding for the installation and support of the testing facility was established by a fee per liter or per gallon on all fuel marketed within the state.

In 1992, the NCWM established the Petroleum Subcommittee under the Laws and Regulations Committee. The Subcommittee recommended major revisions to the Law that was adopted at the 80th NCWM in 1995. The scope of the Law was expanded to include all engine fuels, petroleum products, and automotive lubricants, and its title was changed accordingly. Other changes included expansion of the definitions section, limitation of the scope of the registration section to engine fuels designed for special use, and addition of sections on administrative and civil penalties and on criminal penalties.

In 2007, the Fuel and Lubricants Subcommittee (formerly the Petroleum Subcommittee) undertook a review of this uniform law to update it to eliminate reference to "petroleum products" and reflect the addition of new engine fuels to the marketplace. The amendments included new provisions to provide officials with the authority to review delivery records and grant waivers of requirements adopted under the law in times of emergency or natural disasters.

At the 2008 NCWM Interim Meeting, the Laws and Regulations Committee changed the Petroleum Subcommittee's name to the Fuels and Lubricants Subcommittee (FALS) in recognition of its work with a wide variety of fuels including petroleum and biofuels.

2. Status of Promulgation

The current Uniform Engine Fuels and Automotive Lubricants Inspection Law was recommended for adoption by the Conference in 2008. The table beginning on page 10, Section II. Uniformity of Laws and Regulations of Handbook 130 shows the status of adoption of the law.

(Amended 2008)

**The National Conference on Weights and Measures (NCWM) is supported by the National Institute of Standards and Technology (NIST) in partial implementation of its statutory responsibility for "cooperation with the states in securing uniformity in weights and measures laws and methods of inspection."*

THIS PAGE INTENTIONALLY LEFT BLANK

Uniform Engine Fuels and Automotive Lubricants Inspection Law

Table of Contents

Section **Page**

THIS PAGE INTENTIONALLY LEFT BLANK

Uniform Engine Fuels and Automotive Lubricants Inspection Law

Section 1. Purpose

There should be uniform requirements for engine fuels, non-engine fuels, and automotive lubricants among the states. This Act provides for the establishment of quality specifications for these products.

(Amended 2008)

Section 2. Scope

The Act establishes a sampling, testing, and enforcement program, provides authority for fee collection, requires registration of engine fuels, and empowers the state to promulgate regulations as needed to carry out the provisions of the Act. It also provides for administrative, civil, and criminal penalties.

Section 3. Definitions

As used in this act:

3.1. Engine Fuel. – Any liquid or gaseous matter used for the generation of power in an internal combustion engine.

3.2. Director. – The ______ of the Department of ________ and designated agents.

3.3. Person. – An individual, corporation, company, society, association, partnership, or governmental entity.

3.4. ASTM International. (www.astm.org) – An international voluntary consensus standards organization formed for the development of standards on characteristics and performance of materials, products, systems, and services, and the promotion of related knowledge.

3.5. Automotive Lubricants. – Any material interposed between two surfaces that reduces the friction or wear between them.

3.6. Engine Fuel Designed for Special Use. – Engine fuels designated by the Director requiring registration. These fuels normally have no ASTM or other national consensus standards applying to their quality or usability; common special fuels are racing fuels and those intended for agricultural and other off-road applications.

3.7. Sold. – Kept, offered, or exposed for sale.

3.8. Non-engine Fuels. – Any liquid or gaseous matter used for the generation of heat, power, or similar uses.

(Added 2008)

Section 4. Administration, Adoption of Standards, and Rules

The provisions of this Act shall be administered by the Director. For the purpose of administering and giving effect to the provisions of this Act, the specification and test method standards set forth in the most recent version available of ASTM International standards as published on its website **www.astm.org** are adopted except as amended or modified as required by the Director to comply with federal and state laws. When no ASTM standard exists, other generally recognized national consensus standards may be used. The Director is empowered to write rules and regulations on the advertising, posting of prices, labeling, standards for, and identity of fuels, non-engine fuels, and automotive lubricants and is authorized to establish a testing laboratory.

(Amended 2008)

Section 5. General Duties and Powers

The Director shall have the authority to:

5.1. Enforce and administer all the provisions of this Act by inspections, analyses, and other appropriate actions.

5.2. Have access during normal business hours to all places where engine fuels, non-engine fuels, and automotive lubricants are kept, transferred, offered, exposed for sale, or sold for the purpose of examination, inspection, taking of samples, and review of fuel storage, receipts, transfers, sales records, or delivery records for determining compliance with this Act. If such access is refused by the owner, agent, or other persons leasing the same, the Director may obtain an administrative search warrant from a court of competent jurisdiction.
(Amended 2008)

5.3. Collect, or cause to be collected, samples of engine fuels, non-engine fuels, and automotive lubricants marketed in this state, and cause such samples to be tested or analyzed for compliance with the provisions of this Act.
(Amended 2008)

5.4. Define engine fuels for special use and refuse, revoke, suspend, or issue a stop-order if found not to be in compliance and remand stop-order if the engine fuel for special use is brought into full compliance with this Act.

5.5. Issue a stop-sale order for any engine fuel, non-engine fuels, and automotive lubricant found not to be in compliance and remand a stop-sale order if the engine fuel, petroleum product, or automotive lubricant is brought into full compliance with this Act.
(Amended 2008)

5.6. Refuse, revoke, or suspend the registration of an engine fuel, petroleum product, or automotive lubricant.

5.7. Delegate to appropriate personnel any of these responsibilities for the proper administration of this Act.

5.8. The Director is empowered to waive specific state requirements adopted under this Act or may establish alternative requirements for fuels as determined to be necessary in the event of an emergency or a natural disaster for a specified period of time.
(Added 2008)

Section 6. Registration of Engine Fuels Designed for Special Use

All engine fuels designed for special use must be registered with the Director. Such registration shall include:

6.1. Name, brand, or trademark under which the fuel will be sold.

6.2. Name and address of person registering the engine fuel.

6.3. The special use for which the engine fuel is designed.

6.4. A certification, declaration, or affidavit stating the fuel specifications.

Section 7. Inspection Fee

There shall be a fee of $__________ per appropriate unit of measure on all products covered under the scope of this Act marketed within this state for the purposes of administering and effectively enforcing the provisions of this Act.

Section 8. Prohibited Acts

It shall be unlawful to:

8.1. Represent engine fuels, non-engine fuels, or automotive lubricants in any manner that may deceive or tend to deceive the purchaser as to the nature, brand, price, quantity, and/or quality of such products.
(Amended 1996 and 2008)

8.2. Fail to register an engine fuel designed for special use.

8.3. Submit incorrect, misleading, or false information regarding the registration of an engine fuel designed for special use.

8.4. Hinder or obstruct the Director in the performance of the Director's duties.

8.5. Represent an engine fuel, non-engine fuels, or automotive lubricant that is contrary to the provisions of this Act.
(Amended 2008)

8.6. Represent automotive lubricants with an S.A.E. (Society of Automotive Engineers) viscosity grade or API (American Petroleum Institute) service classification other than those specified by the intended purchaser.
(Added 1996)

Section 9. Civil Penalties

9.1. Assessment of Penalties. – Any person who, by himself or herself, by his or her servant or agent, or as the servant or agent of another person commits any of the acts enumerated in Section 8. Prohibited Acts may be assessed by the __________ a civil penalty of:

(a) not less than $__________ nor more than $__________ for a first violation;

(b) not less than $__________ nor more than $__________ for a second violation within __________ from the date of the first violation; and

(c) not less than $__________ nor more than $__________ for a third violation within __________ from the date of the first violation.

9.2. Administrative Hearing. – Any person subject to a civil penalty shall have a right to request an administrative hearing within __________ days of receipt of the notice of the penalty. The Director or his/her designee shall be authorized to conduct the hearing after giving appropriate notice to the respondent. The decision of the Director shall be subject to appropriate judicial review.

9.3. Collection of Penalties. – If the respondent has exhausted his or her administrative appeals and the civil penalty has been upheld, he or she shall pay the civil penalty within __________ days after the effective date of the final decision. If the respondent fails to pay the penalty, a civil action may be brought by the Director in any court of competent jurisdiction to recover the penalty. Any civil penalty collected under this Act shall be transmitted to __________.

Section 10. Criminal Penalties

10.1. Misdemeanor. – Any person who violates any provision of this Act or regulations promulgated thereto shall be guilty of a Class __________ misdemeanor and upon conviction shall be punished by a fine of not less than $__________ nor more than $__________, or imprisonment for not less than __________ nor more than __________, or both.

10.2. Felony. – Any person who intentionally violates any provision of this Act or regulations promulgated thereto or is convicted under the misdemeanor provisions of this section more than three times in a two-year period shall be guilty of a Class __________ felony and upon conviction shall be punished by a fine of not less than $__________ nor more than $__________, or imprisonment for not less than __________ nor more than __________, or both.

Section 11. Restraining Order and Injunction

The Director is authorized to apply to any court of competent jurisdiction for a restraining order or a temporary or permanent injunction restraining any person from violating any provision of this Act.

Section 12. Severability Provisions

If any word, phrase, provision, or portion of this Act shall be held in a court of competent jurisdiction to be unconstitutional or invalid, the unconstitutionality or invalidity shall apply only to such word, phrase, provision, or portion, and for this purpose the provisions of this Act are declared to be severable.

Section 13. Repeal of Conflicting Laws

All laws and parts of laws contrary to or inconsistent with the provisions of this Act are repealed except as to offense committed, liabilities incurred, and claims made there under prior to the effective date of this Act.

Section 14. Citation

This Act may be cited as the "Engine Fuels and Automotive Lubricants Inspection Act of __________."
(Amended 2008)

Section 15. Effective Date

This Act shall become effective on __________.

IV. Uniform Regulations

Page

THIS PAGE INTENTIONALLY LEFT BLANK

A. Uniform Packaging and Labeling Regulation

as adopted by
The National Conference on Weights and Measures*

1. Background

The Uniform Packaging and Labeling Regulation was first adopted during the 37th Annual Meeting of the National Conference on Weights and Measures (NCWM) in 1952. Reporting to the Conference, the Committee on Legislation stated:

The National Conference should adopt a model package regulation for the guidance of those states authorized to adopt such a regulation under provisions of their weights and measures laws. Since so much of the work of weights and measures officials in the package field concerns food products, the importance of uniformity between the Federal (FDA) regulations and any model regulations to be adopted by this Conference cannot be overemphasized.

Since its inception, the Uniform Packaging and Labeling Regulation has been continually revised to meet the complexities of an enormous expansion in the packaging industry – an expansion that, in late 1966, brought about the passage of the Fair Packaging and Labeling Act (FPLA). Recognizing the need for compatibility with the Federal Act, in 1968 the Committee on Laws and Regulations of the 53rd Annual Meeting of the National Conference amended the "Model Packaging and Labeling Regulation" (renamed in 1983) to parallel regulations adopted by federal agencies under FPLA. The process of amending and revising this Regulation is a continuing one in order to keep it current with practices in the packaging field and make it compatible with appropriate federal regulations. Amendments and additions since 1971 are noted at the end of each section.

The revision of 1978 provided for the use of the metric system (SI) on labels as well as allowing SI-only labels for those commodities not covered by federal laws or regulations. "SI" means the International System of Units as established in 1960 by the General Conference on Weights and Measures and interpreted or modified for the United States by the Secretary of Commerce. [See the "Interpretation of the International System of Units for the United States" in the "Federal Register" (Volume 73, No. 96, pages 28432 to 28433) for May 16, 2008, and 15 United States Code, Section 205a - 205l "Metric Conversion." See also NIST Special Publication 330 "The International System of Units (SI)" 2008 edition and NIST Special Publication 811 "Guide for the Use of the International System of Units (SI)" 2008 edition that are available at **www.nist.gov/pml/wmd/index.cfm** or by contacting TheSI@nist.gov.] In 1988, Congress amended the Metric Conversion Law to declare that it is the policy of the United States to designate the International System of Units of measurement as the preferred system of weights and measures for U.S. trade and commerce. In 1992, Congress amended the federal FPLA to require the most appropriate units of the SI and the customary inch-pound systems of measurement on certain consumer commodities. The 1993 amendments to NIST Handbook 130 require SI and inch-pound units on certain consumer commodities in accordance with federal laws or regulations. Requirements for labeling in both units of measure were effective February 14, 1994, under FPLA and as specified in Section 15 Effective Date; except as specified in Section 11.32. SI Units, Exemptions for Consumer Commodities.

Nothing contained in this regulation should be construed to supersede any labeling requirement specified in federal law or to require the use of SI units on non-consumer packages.

2. Status of Promulgation

The table beginning on page 10, Section II. Uniformity of Laws and Regulations of Handbook 130 shows the status of adoption of the Uniform Packaging and Labeling Regulation.

**The National Conference on Weights and Measures (NCWM) is supported by the National Institute of Standards and Technology (NIST) in partial implementation of its statutory responsibility for "cooperation with the states in securing uniformity in weights and measures laws and methods of inspection."*

THIS PAGE INTENTIONALLY LEFT BLANK

Uniform Packaging and Labeling Regulation

Table of Contents

THIS PAGE INTENTIONALLY LEFT BLANK

Uniform Packaging and Labeling Regulation

Preamble

The purpose of this regulation is to provide accurate and adequate information on packages as to the identity and quantity of contents so that purchasers can make price and quantity comparisons.

(Added 1989)

Section 1. Application

This regulation shall apply to packages, but shall not apply to:

(a) inner wrappings not intended to be individually sold to the customer;

(b) shipping containers or wrapping used solely for the transportation of any commodities in bulk or in quantity to manufacturers, packers, or processors, or to wholesale or retail distributors, but in no event shall this exclusion apply to packages of consumer or non-consumer commodities as defined herein;

(Added 1971)

(c) auxiliary containers or outer wrappings used to deliver packages of such commodities to retail customers if such containers or wrappings bear no printed matter pertaining to any particular commodity;

(d) containers used for retail tray pack displays when the container itself is not intended to be sold (e.g., the tray that is used to display individual envelopes of seasonings, gravies, etc., and the tray itself is not intended to be sold);

(e) open carriers and transparent wrappers or carriers for containers when the wrappers or carriers do not bear any written, printed, or graphic matter obscuring the label information required by this regulation; or

(f) packages intended for export to foreign countries.

(Amended 1994 and 1998)

Section 2. Definitions

2.1. Package. – Except as modified by Section 1. Application, the term "package," whether standard package or random package, means any commodity:

(a) enclosed in a container or wrapped in any manner in advance of wholesale or retail sale; or

(b) whose weight [*NOTE 1*, page 61] or measure has been determined in advance of wholesale or retail sale. An individual item or lot of any commodity on which there is marked a selling price based on an established price per unit of weight or of measure shall be considered a package or packages.

(Amended 1988 and 1991)

NOTE 1: *When used in this regulation, the term "weight" means "mass." (See paragraphs U. "Mass" and "Weight" in Section I. Introduction of NIST Handbook 130 for an explanation of these terms.)*

2.2. Consumer Package of Consumer Commodity. – A package that is customarily produced or distributed for sale through retail sales agencies or instrumentalities for consumption or use by individuals for the purposes of personal care or in the performance of services ordinarily rendered in or about the household or in connection with personal possessions.

(Amended 1988 and 1991)

2.3. Non-consumer Package: Package of Non-consumer Commodity. – Any package other than a consumer package, and particularly a package intended solely for industrial or institutional use or for wholesale distribution.

(Amended 1988 and 1991)

2.4. Random Package. – A package that is one of a lot, shipment, or delivery of packages of the same consumer commodity with no fixed pattern of net contents.

(Amended 1988 and 1990)

2.5. Label. – Any written, printed, or graphic matter affixed to, applied to, attached to, blown into, formed, molded into, embossed on, or appearing upon or adjacent to a consumer commodity, or a package containing any consumer commodity, for purposes of branding, identifying, or giving any information with respect to the commodity or to the contents of the package, except that an inspector's tag or other non-promotional matter affixed to or appearing upon a consumer commodity shall not be considered a label requiring the repetition of label information required by this regulation.

(Amended 1988)

2.6. Person. – The term "person" means either singular or plural and shall include any individual, partnership, company, corporation, association, or society.

(Amended 1988)

2.7. Principal Display Panel or Panels. – That part, or those parts, of a label that is, or are, so designed as to most likely be displayed, presented, shown, or examined under normal and customary conditions of display and purchase. Wherever a principal display panel appears more than once on a package, all requirements pertaining to the "principal display panel" shall pertain to all such "principal display panels."

(Amended 1988)

2.8. Multi-unit Package. – A package containing two or more individual packages of the same commodity, in the same quantity, intended to be sold as a multi-unit package, but where the component packages are labeled individually in full compliance with all requirements of this regulation.

(Amended 1988)

2.9. Combination Package. – A package intended for retail sale, containing two or more individual packages or units of dissimilar commodities.

> **Examples**:
> An antiquing or housecleaning kit;
> sponge and cleaner;
> lighter fluid and flints.

(Added 1989)

2.10. Variety Package. – A package intended for retail sale, containing two or more individual packages or units of similar, but not identical, commodities. Commodities that are generically the same, but that differ in weight, measure, volume, appearance, or quality, are considered similar, but not identical.

> **Examples**:
> Two sponges of different sizes;
> plastic tableware, consisting of 4 spoons, 4 knives, and 4 forks.

(Added 1989)

2.11. Petroleum Products. – Gasoline, diesel fuel, kerosene, or any product (whether or not such a product is actually derived from naturally occurring hydrocarbon mixtures known as "petroleum") commonly used in powering, lubricating, or idling engines or other devices, or is labeled as fuel to power camping stoves or lights.

Therefore, sewing machine lubricant, camping fuels, and synthetic motor oil are "petroleum products" for the purposes of this regulation. Brake fluid, copier machine dispersant, antifreeze, cleaning solvents, and alcohol are not "petroleum products."

(Added 1987) (Amended 1988)

2.12. Spot Label. – A spot label is a label clearly defined by means of a border, indentation, or other means that covers only a small portion of the surface of a principal display panel of a package; the entire portion of the principal display panel outside the area of the label contains no printed or graphic matter of any kind. A spot label may contain all required labeling information (identity, responsibility, and net contents), but it must at least indicate the identity and net contents. See Section 11.29. Spot Label for net contents placement exemption for a spot label.

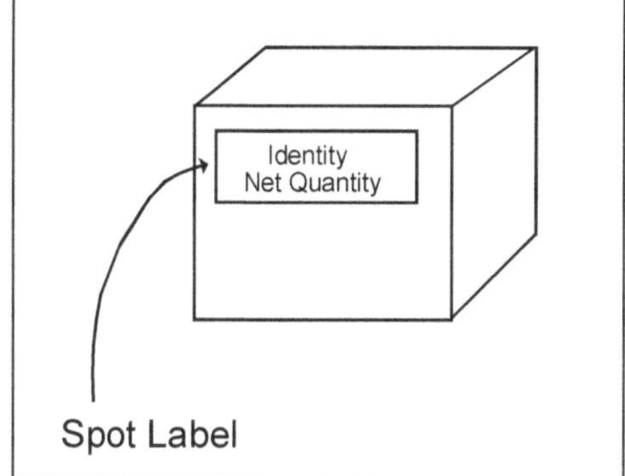

(Added 1990) (Amended 1991)

2.13. Header Strip. – A header label or header strip is a label that is attached across the top of a transparent or opaque bag or other container that bears no other printed or graphic material. See Section 11.30. Header Strip for net contents placement exemptions.

(Added 1990)

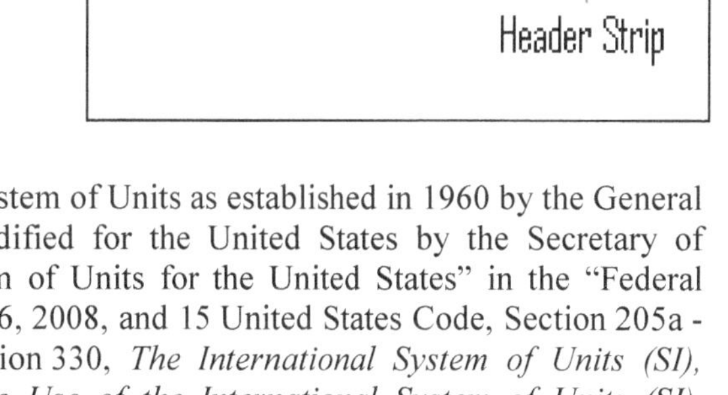

2.14. Standard Package. – A package that is one of a lot, shipment, or delivery of packages of the same commodity with identical net contents declarations.

Examples:
1 L bottles or 12 fl oz cans of carbonated soda,
500 g or 5 lb bags of sugar, or
100 m packages of rope

(Added 1991)

2.15. SI or SI Units. – SI or SI Units means the International System of Units as established in 1960 by the General Conference on Weights and Measures and interpreted or modified for the United States by the Secretary of Commerce. See the "Interpretation of the International System of Units for the United States" in the "Federal Register" (Volume 73, No. 96, pages 28432 to 28433) for May 16, 2008, and 15 United States Code, Section 205a - 2051 "Metric Conversion." See also NIST Special Publication 330, *The International System of Units (SI)*, 2008 edition and NIST Special Publication 811, *Guide for the Use of the International System of Units (SI)*, 2008 edition that are available at **www nist.gov/pml/wmd/index.cfm** or by contacting TheSI@nist.gov.

(Added 1993)

Section 3. Declaration of Identity: Consumer Package

3.1. Declaration of Identity: Consumer Package. – A separate declaration of identity [*NOTE 2*, page 64] on a consumer package shall appear on the principal display panel and shall not be misleading or deceptive. The identity shall be in terms of:

(a) the name specified in or required by any applicable federal or state law or regulation or, in the absence of this;

(b) the common or usual name or, in the absence of this;

(c) the generic name or other appropriate description, including a statement of function (such as "cleaning powder").

(Amended 1990) (Note added 1986)

NOTE 2: *Section 19.(a) of the Uniform Weights and Measures Law (and 21 CFR 101.100 (b) (3) for non-meat and non-poultry foods) specifically exempts food packages from identity statements if the commodity is a food, other than meat or poultry, that was repackaged in a retail establishment and the food is displayed to the purchaser under either of the following circumstances: "(1) its interstate labeling is clearly in view or with a counter card, sign, or other appropriate device bearing prominently and conspicuously the common or usual name of the food, or (2) the common or usual name of the food is clearly revealed by its appearance."*

(Added 1986) (Amended 2001)

3.1.1. Parallel Identity Declaration: Consumer Package. – A declaration of the identity on a consumer package shall appear generally parallel to the base on which the package rests as it is designed to be displayed.

Section 4. Declaration of Identity: Non-consumer Package

A declaration of identity [*NOTE 2*, page 64] on a non-consumer package shall appear on the outside of a package and shall not be misleading or deceptive. The identity shall be in terms of:

(a) the name specified in or required by any applicable federal or state law or regulation or, in the absence of this;

(b) the common or usual name or, in the absence of this;

(c) the generic name or other appropriate description, including a statement of function (such as "cleaning powder").

(Amended 1990) (Note added 1986)

Section 5. Declaration of Responsibility: Consumer and Non-consumer Packages

Any package kept, offered, or exposed for sale, or sold at any place other than on the premises where packed shall specify conspicuously on the label of the package the name and address of the manufacturer, packer, or distributor. The name shall be the actual corporate name, or, when not incorporated, the name under which the business is conducted. The address shall include street address, city, state (or country if outside the United States), and ZIP Code (or the mailing code, if any, used in countries other than the United States); however, the street address may be omitted if this is shown in a current city directory or telephone directory.

If a person manufactures, packs, or distributes a commodity at a place other than his principal place of business, the label may state the principal place of business in lieu of the actual place where the commodity was manufactured or packed or is to be distributed, unless such statement would be misleading. Where the commodity is not manufactured by the person whose name appears on the label, the name shall be qualified by a phrase that reveals the connection such person has with such commodity, such as "Manufactured for and packed by __________," "Distributed by __________," or any other wording of similar import that expresses the facts.

Section 6. Declaration of Quantity: Consumer Packages

6.1. General. [*NOTE 3*, page 64] – The International System of Units (SI), known as the metric system and the inch-pound system of weights and measures are recognized as proper systems to be used in the declaration of quantity. Effective February 14, 1994, appropriate units of both systems shall be presented in a declaration of quantity except as specified in Section 11.32. SI Units, Exemptions for Consumer Commodities and Section 11.33 Inch-Pound Units, Exemptions - Consumer Commodities.

(Amended 1985, 1990, 1993, and 1999)

NOTE 3: *Packages subject to this Section and/or the Federal Fair Packaging and Labeling Act shall be labeled in units of the International System of Units (SI) and the inch-pound system of measure effective February 14, 1994,* *[except for seed (see Section 10.10. Packaged Seed) and camera film and recording tape (see Section 11.22. Camera Film, Video Recording Tape, Audio Recording Tape and Other Image and Audio Recording Media Intended for Retail Sale and Consumer Use), and as specified in Section 11.32. SI Units, Exemptions - Consumer Commodities]. SI units may appear first.*

(Added 1982) (Amended 1990 and 1993)

6.2. Largest Whole Unit. – Where this regulation requires that the quantity declaration be in terms of the largest whole unit, the declaration shall, with respect to a particular package, be in terms of the largest whole unit of weight or measure with any remainder expressed (following the requirements of Section 6.5.2. (a) Fractions and Section 6.11. Fractions):

(a) **SI Units.** – in decimal fractions of such largest whole unit.

(b) **Inch-pound Units.**

(1) in common or decimal fractions of such largest whole unit; or

(2) in the next smaller whole unit or units with any further remainder in terms of common or decimal fractions of the smallest unit present in the quantity declaration.

6.3. Net Quantity. – A declaration of net quantity of the commodity in the package, exclusive of wrappers and any other material packed with such commodity (except as noted in Section 10.3. Aerosols and Similar Pressurized Containers), shall appear on the principal display panel of a consumer package and, unless otherwise specified in this regulation (see Sections 6.6. Prescribed Units, SI, through 6.9. Bi-dimensional Commodities), shall be in terms of the largest whole unit.

6.3.1. Use of "Net Mass" or "Net Weight." – A quantity declaration may stand alone [e.g., "200 g (7 oz)" or "1 lb (453 g)"] or may include the term "net mass" or "net weight" either preceding or following the declaration. The term "net" by itself may be used on food labels. However, the quantity of contents shall always declare the net quantity of contents even when such terms are not used.

(Amended 1993)

6.3.2. Lines of Print or Type. – A declaration of quantity may appear on one or more lines of print or type.

(Amended 1982)

6.4. Terms: Weight, Measure, Volume, or Count. – The declaration of the quantity of a particular commodity shall be expressed in terms of:

(a) weight if the commodity is solid, semisolid, viscous, or a mixture of solid and liquid;

(b) volume measure if the commodity is liquid or dry, if the commodity is dry;

(c) linear measure or area; or

(d) numerical count.

However, if there exists a firmly established general consumer usage and trade custom with respect to the terms used in expressing a declaration of quantity of a particular commodity, such a declaration of quantity may be expressed in its traditional terms, provided such traditional declaration gives accurate and adequate information as to the quantity of the commodity. Any net content statement that does not permit price and quantity comparisons is forbidden.

(Amended 1989)

6.4.1. Combination Declaration.

(a) A declaration of quantity in terms of weight or volume shall be combined with appropriate declarations of the measure, count, and size of the individual units unless a declaration of weight alone is fully informative.

(b) A declaration of quantity in terms of measure shall be combined with appropriate declarations of the weight, volume, count, and size of the individual units unless a declaration of measure alone is fully informative.

(c) A declaration of quantity in terms of count shall be combined with appropriate declarations of the weight, volume, measure, and size of the individual units unless a declaration of count alone is fully informative.

(Added 1971)

6.5. SI Units: Mass, Measure. [*NOTE 3*, page 64] – A declaration of quantity:

(a) in units of mass shall be the kilogram, gram, or milligram;

(b) in units of liquid measure shall be the liter or milliliter and shall express the volume at 20 °C, except in the case of petroleum products or distilled spirits, for which the declaration shall express the volume at 15.6 °C, and except also in the case of a commodity that is normally sold and consumed while frozen, for which the declaration shall express the volume at the frozen temperature, and except also in the case of malt beverages or a commodity that must be maintained in the refrigerated state, for which the declaration shall express the volume at 4 °C;

(Amended 1985 and 1990)

(c) in units of linear measure shall be the meter, centimeter, or millimeter;

(d) in units of area measure shall be the square meter, square decimeters, square centimeter, or square millimeter;

(e) in units of volume other than liquid measure shall be the liter and milliliter, except that the units cubic meter and cubic centimeter shall be used only when specifically designated as a method of sale;

(f) Rule of 1000. – The selected multiple or submultiple prefixes for SI units shall result in numerical values between 1 and 1000. This rule allows centimeters or millimeters to be used where a length declaration is less than 100 centimeters.

Examples:
500 g, not 0.5 kg;
1.96 kg, not 1960 g;
750 mL, not 0.75 L; or
750 mm or 75 cm, not 0.75 m

(Added 1993)

(g) SI declarations should be shown in three digits except where the quantity is below 100 grams, milliliters, centimeters, square centimeters, or cubic centimeters, where it may be shown in two digits. In either case, any final zero appearing to the right of the decimal point need not be shown; and

(Added 1993)

(h) the declaration of net quantity of contents shall not be expressed in mixed units.

Example:
1.5 kg, not 1 kg 500 g.
(Added 1993)

6.5.1. Symbols. – Any of the following symbols for SI units, and none other, may be employed in the quantity statement on a package of commodity:

centimeter	cm	cubic meter	m^3
cubic centimeter	cm^3	kilogram	kg
meter	m	gram	g
milligram	mg	millimeter	mm
liter	L or l	square meter	m^2
milliliter	mL or ml	cubic decimeter	dm^3
square centimeter	cm^2	square decimeter	dm^2
micrometer	µm	microgram	µg or mcg

(a) Symbols [*NOTE 4*, page 67], except for liter, are not capitalized unless the unit is derived from a proper name. Periods shall not be used after the symbol. Symbols shall always be written in the singular form. Adding "s" to an SI symbol to express the plural of the symbol is prohibited.

(b) The "L" symbol and the "mL" symbol are preferred; however, the "l" symbol for liter and "ml" symbol for milliliter are permitted.
(Amended 1980 and 1993)

***NOTE 4**: The "e" mark shall not be considered to be a qualifying word or phrase and may be used as part of the statement of the net quantity of contents where warranted. When used, the "e" mark shall be at least 3 mm (approximately ⅛ in) in height. The term "e" mark refers to the symbol "e" used in connection with the quantity declarations on labels of some consumer commodities marketed primarily in the European Union (EU) and South Africa. The "e" mark constitutes a representation by the packer or importer that the package to which it is applied has been filled in accordance with the average system of quantity specified by the EU. The average system is a method of declaring package fill in the EU and other countries of the world, including the United States.*
(Added 1993)

6.5.2. Fractions and Prefixes.

(a) **Fractions:** An SI statement in a declaration of net quantity of contents of any consumer commodity may contain only decimal fractions.

(b) **Prefixes:** The following chart indicates SI prefixes that may be used on a broad range of consumer commodity labels to form multiples and submultiples of SI units:

Prefix	Symbol	Multiplying Factor*
kilo-	k	x 10^3
deka-**	da	x 10
deci-**	d	x 10^{-1}
centi-***	c	x 10^{-2}
milli-	m	x 10^{-3}
micro-****	µ	x 10^{-6}

*10^2 = 100; 10^3 = 1000; 10^{-1} = 0.1; 10^{-2} = 0.01
Thus, 2 kg = 2 x 1000 g = 2000 g and 3 cm = 3 x 0.01 m = 0.03 m
**Not permitted on food labels.
***Should only be used with "meter."
****Shall only be used for measurements less than 1 mm.

(Amended 1993)

6.6. Prescribed Units, SI. [*NOTE 3*, page 64]

6.6.1. Less than 1 Meter, 1 Square Meter, 1 Kilogram, 1 Cubic Meter, or 1 Liter. – The declaration of quantity shall be expressed as follows:

(a) length measure of less than 1 meter: in centimeters or millimeters;
(Amended 1979)

(b) area measure of less than 1 m^2: in square decimeters and decimal fractions of a square decimeter or in square centimeters and decimal fractions of a square centimeter;

(c) mass of less than 1 kg: in grams and decimal fractions of a gram, but if less than 1 g, then in milligrams;

(d) liquid or dry measure of less than 1 L: in milliliters; and

(e) cubic measure less than 1 m^3: in cubic centimeters or cubic decimeters (liters);
(Added 1993)

provided the quantity declaration appearing on a random mass package may be expressed in units of decimal fractions of the largest appropriate unit, the fraction being carried out to not more than three decimal places.
(Amended 1980 and 1993)

6.6.2. One Meter, 1 Square Meter, 1 Kilogram, 1 Liter, 1 Cubic Meter, or More. – In the case of:

(a) length measure of 1 m or more: in meters and decimal fractions to not more than three places;

(b) area measure of 1 m^2 or more: in square meters and decimal fractions to not more than three places;

(c) mass of 1 kg or more: in kilograms and decimal fractions to not more than three places;

(d) liquid or dry measure of 1 L or more: in liters and decimal fractions to not more than three places; and
(Added 1986) (Amended 1993)

(e) cubic measure of 1 m^3 or more: in cubic meters and decimal fractions to not more than three places.
(Added 1993)

6.7. Inch-Pound Units: Weight, Measure. – A declaration of quantity:

(a) in units of weight shall be in terms of the avoirdupois pound or ounce;

(b) in units of liquid measure shall be in terms of the United States gallon of 231 in^3 or liquid quart, liquid pint, or fluid-ounce subdivisions of the gallon and shall express the volume at 68 °F, except in the case of petroleum products and distilled spirits, for which the declaration shall express the volume at 60 °F, and except also in the case of a commodity that is normally sold and consumed while frozen, for which the declaration shall express the volume at the frozen temperature, and except also in the case of a commodity that must be maintained in the refrigerated state, for which the declaration shall express the volume at 40 °F, and except also in the case of malt beverages, for which the declaration shall express the volume at 39.1 °F;

(Amended 1985 and 1990)

(c) in units of linear measure shall be in terms of the yard, foot, or inch;

(d) in units of area measure shall be in terms of the square yard, square foot, or square inch;

(e) in units of volume measure shall be in terms of the cubic yard, cubic foot, or cubic inch; and

(f) in units of dry measure shall be in terms of the United States bushel of 2150.42 in^3, or peck, dry quart, and dry pint subdivisions of the bushel.

6.7.1. Symbols and Abbreviations. – Any of the following symbols and abbreviations, and none other, shall be employed in the quantity statement on a package of commodity:

avoirdupois	avdp	ounce	oz
piece	pc	count	ct
pint	pt	cubic	cu
pound	lb	each	ea
feet or foot	ft	quart	qt
fluid	fl	square	sq
gallon	gal	weight	wt
inch	in	yard	yd
liquid	liq	drained	dr
diameter	dia		

A period should not be used after the abbreviation. Abbreviations should be written in singular form; and "s" should not be added to express the plural. (For example, "oz" is the symbol for both "ounce" and "ounces.") Both upper and lower case letters are acceptable.

(Added 1974) (Amended 1980, 1990, and 1993)

6.7.2. Units of Two or More Meanings. – When the term "ounce" is employed in a declaration of liquid quantity, the declaration shall identify the particular meaning of the term by the use of the term "fluid;" however, such distinction may be omitted when, by association of terms (for example, as in "1 pint 4 ounces"), the proper meaning is obvious. Whenever the declaration of quantity is in terms of the dry pint or dry quart, the declaration shall include the word "dry."

(Amended 1982)

6.8. Prescribed Units, Inch-pound System.

6.8.1. Less than 1 foot, 1 square foot, 1 pound, or 1 pint. – The declaration of quantity shall be expressed in the following terms:

(a) in the case of length measure of less than 1 ft, in inches and fractions of inches;

(b) in the case of area measure of less than 1 ft^2, in square inches and fractions of square inches;

(c) in the case of weight of less than 1 lb, in ounces and fractions of ounces; and

(d) in the case of liquid measure of less than 1 pt, in fluid ounces and fractions of fluid ounces, provided, the quantity declaration appearing on a random package may be expressed in terms of decimal fractions of the largest appropriate unit, the fraction being carried out to not more than three decimal places.

(Amended 1984)

6.8.2. One Foot, 1 Square Foot, 1 Pound, 1 Pint, 1 Gallon, or More. – The declaration of quantity shall be expressed in the following terms (see Section 6.2. Largest Whole Unit and Section 6.11. Fractions):

(a) **Linear Measure.** – If 1 ft or more, expressed in terms of the largest whole unit (a yard or a foot) with any remainder expressed in inches and fractions of the inch or in fractions of the foot or yard, except that it shall be optional to include a statement of length in terms of inches.

(b) **Area Measure.**

(1) If 1 ft^2 or more, but less than 4 ft^2, expressed in square feet with any remainder expressed in square inches and fractions of a square inch or in fractions of a square foot; and

(2) If 4 ft^2 or more, expressed in terms of the largest whole unit (e.g., square yards or square feet) with any remainder expressed in square inches and fractions of a square inch or in fractions of the square foot or square yard.

(c) **Weight.** – If 1 lb or more, expressed in terms of the largest whole unit with any remainder expressed in ounces and fractions of an ounce or in fractions of the pound.

(d) **Liquid Volume.**

(1) If 1 pt or more, but less than 1 gal, expressed in the largest whole unit (quarts, quarts and pints, or pints, as appropriate) with any remainder expressed in fluid ounces or fractions of the pint or quart, except that 2 qt may be declared as ½ gal, and it shall be optional to include an additional expression of net quantity in fluid ounces; or

(2) If 1 gal or more, expressed in terms of the largest whole unit (gallons followed by fractions of a gallon or by the next smaller whole unit or units [for example, quarts and pints]) with any remainder expressed in fluid ounces or fractions of the pint or quart, except that it shall be optional to include an additional expression of net quantity in fluid ounces.

(e) **Dry Measure.** – If 1 dry pt or more, expressed in terms of the largest whole unit with the remainder expressed in fractions of a dry pint, dry quart, peck, or bushel, provided the quantity declaration on a random package may be expressed in decimal fractions of the largest appropriate unit carried out to not more than three decimal places.

(Amended 1993)

6.9. Bi-dimensional Commodities. – For bi-dimensional commodities (including roll-type commodities) the quantity declaration shall be expressed in both SI and inch-pound units of measurement as follows:

(a) if the area is less than 929 cm^2 (1 ft^2), in terms of length and width (expressed in the largest whole unit for SI and in linear inches and fractions of linear inches for inch-pound);

Example:
20.3 cm x 25.4 cm (8 in x 10 in);

(b) if the area is at least 929 cm^2 (1 ft^2), but less than 37.1 dm^2 (4 ft^2), in terms of area (expressed in the largest whole unit for SI and in square inches for inch-pound), followed by a declaration of the length and width in terms of the largest whole unit:

Example:
31 dm^2 (49 cm x 64 cm) 3.36 ft^2 (1.6 ft x 2.1 ft), provided:

(1) bi-dimensional commodities having a width of 10 cm (4 in) or less, the declaration of net quantity shall be expressed in terms of width and length in linear measure; no declaration of area is required;

(2) an inch-pound dimension of less than 2 ft may be stated in inches;

(3) commodities consisting of usable individual units (e.g., paper napkins) require a declaration of unit area but not a declaration of total area of all such units (except roll-type commodities with individual usable units created by perforations, for which see Section 6.10. Count: Ply); and

(4) inch-pound declarations may include after the statement of the linear dimensions in the largest whole unit a parenthetical declaration of the same dimensions in inches.

Example:
25 ft^2 (12 in x 8.33 yd) (12 in x 300 in).

(c) if the area is 37.1 dm^2 (4 ft^2) or more, in terms of area (expressed in the largest whole unit for SI and in square feet for inch-pound), followed by a declaration of the length and width, in terms of the largest whole unit, provided:

(1) no declaration of area is required for a bi-dimensional commodity with a width of 10 cm (4 in) or less;

(2) bi-dimensional commodities with a width of 10 cm (4 in) or less, the inch-pound statement of width shall be expressed in terms of linear inches and fractions thereof, and length shall be expressed in the largest whole unit (yard or foot) with any remainder in terms of fractions of the yard or foot, except that it shall be optional to express the length in the largest whole unit followed by a statement of length in inches or to express the length in inches followed by a statement of length in the largest whole unit;

Examples:
5 cm x 9.14 m (2 in x 10 yd); or
5 cm x 9.14 m (2 in x 10 yd) (360 in); or
5 cm x 9.14 m (2 in x 360 in) (10 yd).

(3) an inch-pound dimension of less than 2 ft may be stated in inches; and

(d) no declaration of area is required for commodities for which the length and width measurements are critical in terms of end use (such as wallpaper border) if such commodities clearly present the length and width measurements on the label.

6.10. Count: Ply. – If the commodity is in individually usable units of one or more components or plies, the quantity declaration shall, in addition to complying with other applicable quantity declaration requirements of this regulation, include the number of plies and total number of usable units.

Roll type commodities, when perforated so as to identify individual usable units, shall not be deemed to be made up of usable units; however, such roll type commodities shall be labeled in terms of:

(a) total area measurement;

(b) number of plies;

(c) count of usable units; and

(d) dimensions of a single usable unit.
(Amended 1988)

6.11. Fractions.

(a) **Inch-pound:** An inch-pound statement of net quantity of contents of any consumer commodity may contain common or decimal fractions. A common fraction shall be in terms of halves, quarters, eighths, sixteenths, or thirty-seconds, except that:

(1) if there exists a firmly established general consumer usage and trade custom of employing different common fractions in the net quantity declaration of a particular commodity, they may be employed; and

(2) if linear measurements are required in terms of yards or feet, common fractions may be in terms of thirds.

(b) **Common fractions:** A common fraction shall be reduced to its lowest term.

Example: $^2/_4$ becomes ½

(c) **Decimal fractions:** A decimal fraction shall not be carried out to more than three places.
(Amended 1986 and 1993)

6.12. Supplementary Quantity Declarations. – The required quantity declaration may be supplemented by one or more declarations of weight, measure, or count, such declaration appearing other than on a principal display panel. Such supplemental statement of quantity of contents shall not include any term qualifying a unit of weight, measure, or count that tends to exaggerate the amount of commodity contained in the package (e.g., "giant" quart, "larger" liter, "full" gallon, "when packed," "minimum," or words of similar import).

6.13. Rounding. [*NOTE 5*, page 72] – In all conversions for the purpose of showing an equivalent SI or inch-pound quantity to a rounded inch-pound or SI quantity, or in calculated values to be declared in the net quantity statement, the number of significant digits retained must be such that accuracy is neither sacrificed nor exaggerated. Conversions, the proper use of significant digits, and rounding must be based on the packer's knowledge of the accuracy of the original measurement that is being converted. In no case shall rounded net contents declarations overstate a quantity; the packer may round converted values down to avoid overstating the net contents.
(Amended 1993)

NOTE 5: *When as a result of rounding SI or customary inch-pound declarations the resulting declarations are not exact, the largest declaration (either metric or inch-pound) will be used for enforcement purposes to determine whether a package contains at least the declared amount of the product.*

6.14. Qualification of Declaration Prohibited. – In no case shall any declaration of quantity be qualified by the addition of the words "when packed," "minimum," or "not less than" or any words of similar import (e.g., "approximately"), nor shall any unit of weight, measure, or count be qualified by any term (such as "jumbo," "giant," "full," or the like) that tends to exaggerate the amount of commodity.
(Amended 1998)

6.15. Character of Declaration: Average. – The average quantity of contents in the packages of a particular lot, shipment, or delivery shall at least equal the declared quantity, and no unreasonable shortage in any package shall be

permitted even though overages in other packages in the same shipment, delivery, or lot compensate for such shortage.
(Added 1981)

6.16. Random Packages. – A random weight package must bear a label conspicuously declaring:

(a) the net weight;

(b) unit price; and

(c) the total price.

In the case of a random package packed at one place for subsequent sale at another, neither the price per unit of weight nor the total selling price need appear on the package, provided the package label includes both such prices at the time it is offered or exposed for sale at retail.
(Added 1999)

Section 7. Declaration of Quantity: Non-consumer Packages

7.1. General. – The SI and inch-pound systems of weights and measures are recognized as proper systems to be used in the declaration of quantity. Units of both systems may be combined in a dual declaration of quantity. [*NOTE 6*, page 73] (See Section 6.3. Net Quantity, and Section 6.3.1. Use of "Net Mass" or "Net Weight.")

NOTE 6: Although non-consumer packages under this Regulation may bear SI declarations only, this Regulation should not be construed to supersede any labeling requirement specified in federal law.

7.2. Location. – A non-consumer package shall bear on the outside a declaration of the net quantity of contents. Such declaration shall be in terms of the largest whole unit (see Section 6.2. Largest Whole Unit; for small packages, see Section 11.16. Small Packages).

7.3. Terms: Weight, Liquid Measure, Dry Measure, or Count. – The declaration of the quantity of a particular commodity shall be expressed in terms of liquid measure if the commodity is liquid, in terms of dry measure if the commodity is dry, in terms of weight if the commodity is solid, semisolid, viscous, or a mixture of solid and liquid, or in terms of numerical count. However, if there exists a firmly established general consumer usage and trade custom with respect to the terms used in expressing a declaration of quantity of a particular commodity, such declaration of quantity may be expressed in its traditional terms if such traditional declaration gives accurate and adequate information as to the quantity of the commodity.

7.4. SI Units: Mass, Measure. – A declaration of quantity:

(a) in units of mass shall be in terms of the kilogram, gram, or milligram;

(b) in units of liquid measure shall be in terms of the liter or milliliter, and shall express the volume at 20 °C, except in the case of petroleum products or distilled spirits, for which the declaration shall express the volume at 15.6 °C, and except also in the case of a commodity that is normally sold and consumed while frozen, for which the declaration shall express the volume at the frozen temperature, and except also in the case of malt beverages or a commodity that is normally sold in the refrigerated state, for which the declaration shall express the volume at 4 °C;
(Amended 1985)

(c) in units of linear measure shall be in terms of the meter, centimeter, or millimeter;

(d) in units of area measure shall be in terms of the square meter, square decimeter, square centimeter or square millimeter;

(e) in units of volume other than liquid measure shall be in terms of the liter and milliliter, except that the terms cubic meter, cubic decimeter, and cubic centimeter will be used only when specifically designated as a method of sale;

(f) Rule of 1000. – The selected multiple or submultiple prefixes for SI units shall result in numerical values between 1 and 1000. This rule allows centimeters or millimeters to be used where a length declaration is less than 100 centimeters;

Examples:
500 g, not 0.5 kg;
1.96 kg, not 1960 g;
750 mL, not 0.75 L; or
750 mm or 75 cm, not 0.75 m;

(Added 1993)

(g) SI declarations should be shown in three digits except where the quantity is below 100 grams, milliliters, centimeters, square centimeters, or cubic centimeters where it can be shown in two digits. In either case, any final zero appearing to the right of the decimal point need not be shown; and

(Added 1993)

(h) the declaration of net quantity of contents shall not be expressed in mixed units.

Example:
1.5 kg, not 1 kg 500 g

7.4.1. Symbols. – Only those symbols as detailed in Section 6.5.1. Symbols, and none other, may be employed in the quantity statement on a package of commodity.

7.5. Inch-pound Units: Weight, Measure. – A declaration of quantity:

(a) in units of weight shall be in terms of the avoirdupois pound or ounce;

(b) in units of liquid measure shall be in terms of the United States gallon of 231 cubic inches or liquid quart, liquid pint, or fluid ounce subdivisions of the gallon and shall express the volume at 68 °F, except in the case of petroleum products or distilled spirits, for which the declaration shall express the volume at 60 °F, and except also in the case of a commodity that is normally sold and consumed while frozen, for which the declaration shall express the volume at the frozen temperature, and except also in the case of a commodity that is normally sold in the refrigerated state, for which the declaration shall express the volume at 40 °F, and except also in the case of malt beverages, for which the declaration shall express the volume at 39.1 °F;

(Amended 1985)

(c) in units of linear measure shall be in terms of the yard, foot, or inch;

(d) in units of area measure shall be in terms of the square yard, square foot, or square inch;

(e) in units of volume measure shall be in terms of the cubic yard, cubic foot, or cubic inch; and

(f) in units of dry measure, shall be in terms of the United States bushel of 2150.42 in^3, or peck, dry quart, and dry pint subdivisions of the bushel.

7.5.1. Symbols and Abbreviations. – Any generally accepted symbol and abbreviation of a unit name may be employed in the quantity statement on a package of commodity. (For commonly accepted symbols and abbreviations, see Section 6.7.1. Symbols and Abbreviations.)

7.6. Character of Declaration: Average. – The average quantity of contents in the packages of a particular lot, shipment, or delivery shall at least equal the declared quantity, and no unreasonable shortage in any package shall be permitted, even though overages in other packages in the same shipment, delivery, or lot compensate for such shortage.

Section 8. Prominence and Placement: Consumer Packages

8.1. General. – All information required to appear on a consumer package shall appear thereon in the English language and shall be prominent, definite, plain, and conspicuous as to size and style of letters and numbers and as to color of letters and numbers in contrast to color of background. Any required information that is either in hand lettering or hand script shall be entirely clear and equal to printing in legibility.

8.1.1. Location. – The declaration or declarations of quantity of the contents of a package shall appear in the bottom 30 % of the principal display panel or panels. For cylindrical containers, see also Section 10.7. Cylindrical Containers for additional requirements. For small packages, see Section 11.16. Small Packages.

(Amended 1975)

8.1.2. Style of Type or Lettering. – The declaration or declarations of quantity shall be in such a style of type or lettering as to be boldly, clearly, and conspicuously presented with respect to other type, lettering, or graphic material on the package, except that a declaration of net quantity blown, formed, or molded on a glass or plastic surface is permissible when all label information is blown, formed, or molded on the surface.

8.1.3. Color Contrast. – The declaration or declarations of quantity shall be in a color that contrasts conspicuously with its background, except that a declaration of net quantity blown, formed, or molded on a glass or plastic surface shall not be required to be presented in a contrasting color if no required label information is on the surface in a contrasting color.

8.1.4. Free Area. – The area surrounding the quantity declaration shall be free of printed information:

(a) above and below, by a space equal to at least the height of the lettering in the declaration; and

(b) to the left and right, by a space equal to twice the width of the letter "N" of the style and size of type used in the declaration.

8.1.5. Parallel Quantity Declaration. – The quantity declaration shall be presented in such a manner as to be generally parallel to the declaration of identity and to the base on which the package rests as it is designed to be displayed.

8.2. Calculation of Area of Principal Display Panel for Purposes of Type Size. – The area of the principal display panel shall be:

(a) in the case of a rectangular container, one entire side that properly can be considered to be the principal display panel, the product of the height times the width of that side;

For Figure 3, the area of the principal display panel is 20 cm (8 in) x 15 cm (6 in) = 300 cm^2 (48 in^2).

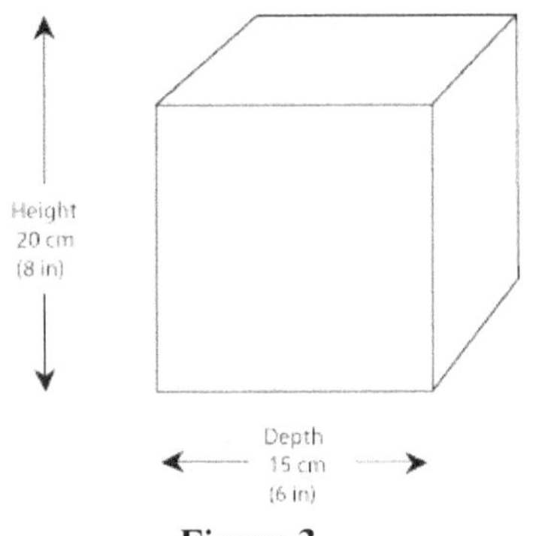

Figure 3.

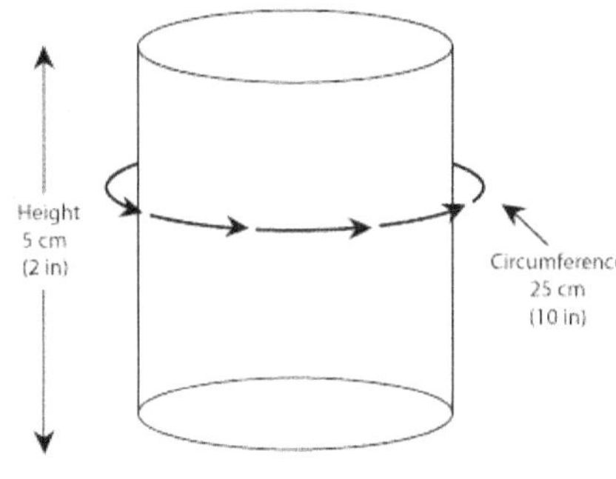

Figure 4.

(b) in the case of a cylindrical or nearly cylindrical container, 40 % of the product of the height of the container times the circumference;

For Figure 4, the area of the principal display panel is:
25 cm (10 in) x 5 cm (2 in) = 125 cm (20 in^2) x 0.40 = 50 cm^2 (8 in^2)
(see also Section 10.7. Cylindrical Containers).

The area of the principal display panel is the same in both examples. The declaration of net quantity of contents must be of the same height in both cases. It is not the size of the label that is used to determine the minimum type size of the quantity statement, but the size of the surface of the package exposed to view to the customer. The package on the right side of the figure has a spot label (see Section 2.12. Spot Label and Section 11.29. Spot Label); and

(c) in the case of any other shaped container, 40 % of the total surface of the container, unless such container presents an obvious principal display panel (e.g., the top of a triangular or circular package of cheese, or the top of a can of shoe polish), in which event the area shall consist of the entire such surface.

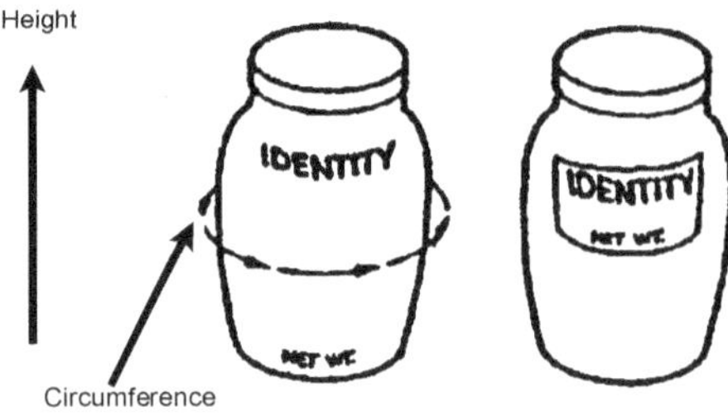

Figure 5.

Determination of the principal display panel shall exclude tops, bottoms, flanges at tops and bottoms of cans, and shoulders and necks of bottles or jars.

8.2.1. Minimum Height of Numbers and Letters. – The height of any letter or number in the required quantity declaration shall be not less than that shown in Table 1 with respect to the area of the panel, and the height of each number of a common fraction shall meet one-half the minimum height standards. When upper and lowercase or all lowercase letters are used, it is the lowercase "o" or its equivalent that shall meet the minimum height requirement. When upper and lowercase or all lowercase letters are used in SI symbols, it is the uppercase "L," lowercase "d," or their equivalent in the print or type that shall meet the minimum height requirement. However, no letter shall be less than 1.6 mm ($^{1}/_{16}$ in) in height. Other letters and exponents must be presented in the same type style and in proportion to the type size used.

(Amended 1993)

8.2.2. Numbers and Letters: Proportion. – No number or letter shall be more than three times as high as it is wide.

Table 1. Minimum Height of Numbers and Letters

Area of Principal Display Panel	Minimum Height of Numbers and Letters	Minimum Height; Label Information Blown, Formed, or Molded on Surface of Container
≤ 32 cm^2 (5 in^2)	1.6 mm (1/16 in)	3.2 mm (1/8 in)
> 32 cm^2 (5 in^2) ≤ 161 cm^2 (25 in^2)	3.2 mm (1/8 in)	4.8 mm (3/16 in)
> 161 cm^2 (25 in^2) ≤ 645 cm^2 (100 in^2)	4.8 mm (3/16 in)	6.4 mm (¼ in)
> 645 cm^2 (100 in^2) 2581 cm^2 (400 in^2)	6.4 mm (¼ in)	7.9 mm (5/16 in)
> 2581 cm^2 (400 in^2)	12.7 mm (½ in)	14.3 mm (9/16 in)

Symbols: ≤ means less than or equal to; < means less than; > means greater than.
NOTE: The type size requirements specified in this table do not apply to the "e" mark [*NOTE 4*, page 67].

Section 9. Prominence and Placement: Non-consumer Packages

9.1. General. – All information required to appear on a non-consumer package shall be definitely and clearly stated thereon in the English language. Any required information that is either in hand lettering or hand script shall be entirely clear and equal to printing in legibility.

Section 10. Requirements: Specific Consumer Commodities, Non-consumer Commodities, Packages, Containers

(Title amended 1979)

10.1. Display Card Package. – For an individual package affixed to a display card, or for a commodity and display card together comprising a package, the type size of the quantity declaration is governed by the dimensions of the display card.

10.2. Eggs. – When cartons containing 12 eggs have been designed so as to permit division in half by the retail purchaser, the required quantity declaration shall be so positioned as to have its context destroyed when the carton is divided.

10.3. Aerosols and Similar Pressurized Containers. – The declaration of quantity on an aerosol package and on a similar pressurized package shall disclose the net quantity of the commodity (including propellant), in terms of weight, that will be expelled when the instructions for use as shown on the container are followed.

10.4. Multi-unit Packages. [*NOTE 7*, page 78] – Any package containing more than one individual "commodity in package form" (see Section 2.1. Package) of the same commodity shall bear on the outside of the package a declaration of:

(a) the number of individual units;

(b) the quantity of each individual unit; and

(c) the total quantity of the contents of the multi-unit package.

Example:
Soap bars, 6 Bars, Net Wt 100 g (3.53 oz) each
Total Net Wt 600 g (1.32 lb).

The term "total" or the phrase "total contents" may precede the quantity declaration.

A multi-unit package containing unlabeled individual packages which are not intended for retail sale separate from the multi-unit package may contain, in lieu of the requirements of section (a), a declaration of quantity of contents expressing the total quantity of the multi-unit package without regard for inner packaging. For such multi-unit packages it shall be optional to include a statement of the number of individual packages when such a statement is not otherwise required by the regulations.

Examples:
Deodorant Cakes:
5 Cakes, Net Wt 113 g (4 oz) each, Total Net Wt 566 g (1.25 lb); or
5 Cakes, Total Net Wt 566 g (1 lb 4 oz)

Soap Packets:
10 Packets, Net Wt 56.6 g (2 oz) each, Total Net Wt 566 g (1.25 lb); or Net Wt 566 g (1 lb 4 oz); or
10 Packets, Total Net Wt 566 g (1 lb 4 oz)

(Amended 1993)

NOTE 7: *For foods, a "multi-unit" package means a package containing two or more individually packaged units of the identical commodity in the same quantity, intended to be sold as part of the multi-unit package but labeled to be individually sold in full compliance with this regulation. Open multi-unit retail food packages under the authority of the FDA or the USDA that do not obscure the number of units or prevent examination of the labeling on each of the individual units are not required to declare the number of individual units or the total quantity of contents of the multi-unit package if the labeling of each individual unit complies with requirements so that it is capable of being sold individually. (See also Section 11.11. Soft Drink Bottles and Section 11.12. Multi-Unit Soft-Drink Bottles.)*

(Added 1984)

10.5. Combination Packages. – A combination package is a package intended for retail sale, containing two or more individual packages or units of dissimilar commodities. The declaration of net quantity for a combination package shall contain an expression of weight, volume, measure, or count or a combination thereof, as appropriate, for each individual package or unit, provided the quantity statements for identical packages or units shall be combined. This section does not apply to food or other commodities subject to the Federal Food, Drug, and Cosmetic Act (21 USC).

Examples:
Lighter Fluid and Flints –
2 cans lighter fluid – each 236 mL (8 fl oz)
1 package – 8 flints

Sponges and Cleaner –
2 sponges – each 10 cm x 15 cm x 2.5 cm (4 in x 6 in x 1 in)
1 box cleaner – Net Mass 170 g (6 oz)

Picnic Pack –
20 spoons, 10 knives, and 10 forks
10 2-ply napkins 25 cm x 25 cm (10 in x 10 in)
10 cups – 177 mL (6 fl oz)

(Amended 1993)

10.6. Variety Packages. – A variety package is a package intended for retail sale, containing two or more individual packages or units of similar but not identical commodities. Commodities that are generically the same but that differ in weight, measure, volume, appearance, or quality are considered similar but not identical. This section does not apply to foods or other commodities subject to the Federal Food, Drug, and Cosmetic Act (21 USC). The declaration of net quantity for a variety package will be expressed as follows:

(a) the number of units for each identical commodity followed by the weight, volume, or measure of that commodity;

(b) the total quantity by weight, volume, measure, and count, as appropriate, of the variety package. The statement of total quantity shall appear as the last item in the declaration of net quantity and shall not be of greater prominence than other terms used.

Examples:
Sponges –
11 Sponges 11 cm x 20.3 cm x 1.9 cm (4 in x 8 in x ¾ in)
14 Sponges 5.7 cm x 10 cm x 1.2 cm (2¼ in x 4 in x ½ in)
Total: 25 Sponges

Soap –
2 Soap Bars 85 g (3 oz) ea
1 Soap Bar 142 g (5 oz)
Total: 3 Soap Bars 312 g (11 oz)

Liquid Shoe Polish –
1 Brown 89 mL (3 fl oz)
1 Black 89 mL (3 fl oz)
1 White 148 mL (5 fl oz)
Total: 326 mL (11 fl oz)

Picnic Ware –
34 spoons
33 forks
33 knives
Total: 100 pieces

(Amended 1993)

When individual units in a variety package are either packaged or labeled and are intended for retail sale as individual units, each unit shall be labeled in compliance with the applicable sections of this regulation.

10.7. Cylindrical Containers. – In the case of cylindrical or nearly cylindrical containers, information required to appear on the principal display panel shall appear within that 40 % of the circumference which is most likely to be displayed, presented, shown, or examined under customary conditions of display for retail sale.

10.8. Measurement of Container-Type Commodities, How Expressed.

10.8.1. General. – Commodities designated and sold at retail to be used as containers for other materials or objects, such as bags, cups, boxes, and pans, shall be labeled with the declaration of net quantity as follows:

(a) For bag-type commodities, in terms of count followed by linear dimensions of the bag (whether packaged in a perforated roll or otherwise). The linear dimensions shall be expressed:

(1) in SI units: in millimeters or centimeters, except that a dimension of 1 meter or more will be expressed in meters with the remainder in terms of decimal fractions of the meter; and

(2) in inch-pound units: in inches, except that a dimension of 2 ft or more will be expressed in feet with any remainder in terms of inches or common or decimal fractions of the foot.

(b) When the unit bag is characterized by two dimensions because of the absence of a gusset, the width and length will be stated.

Examples:
25 Bags, 12.7 cm x 10 cm (5 in x 4 in) or
50 Bags, 75 cm x 1.2 m (2.5 ft x 3.9 ft)

(c) When the unit bag is gusseted, the dimensions will be expressed as width, depth, and length.

Examples:
25 Bags, 43 cm x 10 cm x 50 cm (17 in x 4 in x 20 in) or
100 Bags, 50.8 cm x 30.4 cm x 76.2 cm (20 in x 12 in x 2½ ft)

(d) For other square, oblong, rectangular, or similarly shaped containers, in terms of count followed by length, width, and depth, except depth need not be listed when less than 5 cm or 2 in. The linear dimensions shall be expressed as specified in Section 10.8.1.(a).

Example:
bag-type commodities: 2 Pans, 20 cm x 20 cm (8 in x 8 in)

(e) For circular or other generally round-shaped containers, except cups and the like, in terms of count followed by diameter and depth, except depth need not be listed when less than 5 cm or 2 in.

Example:
4 Pans, 20 cm (8 in) diameter x 10 cm (4 in)

(f) Cups – Notwithstanding the above requirements, the net quantity statement for containers such as cups will be listed in terms of count and liquid capacity per unit.

Example:
24 Cups, 177 mL (6 fl oz) capacity

10.8.2. Capacity. – When the functional use of the container is related by label references in standard terms of measure to the capability of holding a specific quantity of substance or class of substances such references shall be a part of the net quantity statement and shall specify capacity as follows:

(a) in SI units: in terms of volume for all containers and liners. The expressed capacity will be stated in terms of milliliters, except that a quantity of 1 liter or more shall be expressed in liters with the remainder in terms of decimal fractions of the liter; and

(b) in inch-pound units:

(1) In terms of liquid measure for containers that are intended to be used for liquids, semisolids, viscous materials, or mixtures of solids and liquids. The expressed capacity will be stated in terms of the largest whole unit (gallon, quart, pint, fluid ounce) with any remainder in terms of common or decimal fractions of that unit.

Example:
Freezer Boxes – 4 Boxes, 946 mL capacity, 15 cm x 15 cm x 10 cm (1 qt capacity, 6 in x 6 in x 4 in)

(2) In terms of dry measure for containers that are intended to be used for solids. The expressed capacity will be stated in terms of the largest whole unit (bushel, peck) with any remainder in terms of common or decimal fractions of that unit.

Example:
Leaf Bags – 8 Bags, 211 L capacity, 1.21 m x 1.52 m (6 bu capacity, 4 ft x 5 ft)

(3) Where containers are used as liners for other more permanent containers, in the same terms as are normally used to express the capacity of the more permanent containers.

Example:
Garbage Can Liners – 10 Liners, 76.2 cm x 93.9 cm, fits up to 113 L cans (2 ft 6 in x 3 ft 1 in, fits up to 30 gal cans)

10.8.3. Terms. – For purposes of this section, the use of the terms "capacity," "diameter," and "fluid" is optional.

10.9. Textile Products, Threads, and Yarns.

10.9.1. Wearing Apparel. – Wearing apparel (including non-textile apparel and accessories such as leather goods and footwear) sold as single unit items, or if normally sold in pairs (such as hosiery, gloves, and shoes) sold as single unit pairs, shall be exempt from the requirements for net quantity statement by count, as required by Section 6.4. Terms: Weight, Liquid Measure, Dry Measure, or Count of this regulation.

10.9.2. Textiles. – Bed sheets, blankets, pillowcases, comforters, quilts, bedspreads, mattress covers and pads, afghans, throws, dresser and other furniture scarves, tablecloths and napkins, flags, curtains, drapes, dishtowels, dishcloths, towels, facecloths, utility cloths, bathmats, carpets and rugs, potholders, fixture and appliance covers, nonrectangular diapers, slipcovers, etc., shall be exempt from the requirements of Section 6.9. Bi-dimensional Commodities of this regulation, provided:

(a) The quantity statement for fitted sheets and mattress covers shall state, in centimeters and inches, the length and width of the mattress for which the item is designed and the size designation of the mattress if the item is intended to fit a mattress identified as "twin," "double," "queen," "king," "California king," etc.

Example:
Double sheet for 137 cm x 190 cm (54 in x 75 in) mattress.

(Amended 1987)

(b) The quantity statement for flat sheets shall state, in centimeters and inches, the length and width of the mattress for which the sheet is designed, followed in parentheses by a statement, in centimeters and inches, of the length and width of the finished sheet. The quantity statement shall also state the size designation of the mattress for which the sheet is designed, such as "twin," "double," "queen," "king," "California king," if the item is intended to fit such a mattress.

Example:
Twin Flat Sheet for 99 cm x 190 cm (39 in x 75 in) mattress 167 cm x 244 cm (66 in x 96 in) finished size.

(Amended 1987)

(c) The quantity statement for pillowcases shall state, in centimeters and inches, the length and width of the pillow for which the pillowcase is designed, followed in parentheses by a statement, in centimeters and inches of the length and width of the finished pillowcase. The quantity statement for pillowcases shall also state the size designation of the pillow for which the pillowcase is designed, e.g., "youth," "standard," "queen," etc., if the item is intended to fit such pillows.

Example:
Standard Pillowcase for 51 cm x 66 cm (20 in x 26 in) pillow, 51 cm x 76 cm (20 in x 30 in) in finished size.

(Amended 1977 and 1987)

(d) The quantity statement for blankets, comforters, quilts, bedspreads, mattress pads, afghans, and throws shall state, in centimeters and inches, the length and width of the finished item. The quantity statement shall also state the length of any ornamentation and the size designation of the mattress for which the item is designed, if it is intended to fit, for example, a "twin," "double," "queen," "king," "California king," etc., mattress.

(Amended 1988)

(e) The quantity statement for tablecloths and napkins shall state, in centimeters and inches, the length and width of the finished item. The quantity statement also may state parenthetically, in centimeters and inches, the length and width of the item before hemming and be properly identified as such.

(f) The quantity statement for curtains, drapes, flags, furniture scarves, etc., shall state, in centimeters and inches, the length and width of the finished item. The quantity statement also may state parenthetically, in centimeters and inches, the length of any ornamentation.

(g) The quantity statement for carpets and rugs shall state, in meters and feet, with any remainder in decimal fractions of the meter for SI sizes or common or decimal fractions of the foot or in inches for inch-pound sizes, the length and width of the item. The quantity statement also may state parenthetically, in centimeters and inches, the length of any ornamentation.

(h) The quantity statement for woven dishtowels, dishcloths, towels, facecloths, utility cloths, bathmats, etc., shall state, in centimeters and inches, the length and width of the item. The quantity statement for such items, when knitted, need not state the dimensions.

(i) The quantity statement for textile products such as potholders, fixture and appliance covers, slipcovers, non-rectangular diapers, etc., shall be stated in terms of count and may include size designations and dimensions.

(j) The quantity statement for other than rectangular textile products identified in Sections (a) through (h) shall state the geometric shape of the product and the dimensions that are customarily used in describing such geometric shape.

Examples:
Round Scarf 190 cm (74 in) in diameter;
Oval Tablecloth 177 cm x 254 cm (70 in x 100 in) representing the maximum length and width in this case.

(k) The quantity statement for packages of remnants of textile products of assorted sizes, when sold by count, shall be accompanied by the term "irregular dimensions" and the minimum size of such remnants.

(Added 1971)

10.9.3. Sewing Threads, Handicraft Threads, and Yarns. – Sewing and handicraft threads shall be labeled as follows:

(a) The net quantity for sewing and handicraft threads shall be expressed in terms of meters and yards.

(b) The net quantity statement for yarns shall be expressed in terms of mass or weight.

(c) Thread products may, in lieu of name and address, bear a trademark, symbol, brand, or other mark that positively identifies the manufacturer, packer, or distributor provided such marks are filed with the Director.

(d) Each unit of industrial thread shall be marked to show its net length in terms of meters and yards or its net weight in terms of kilograms or grams and avoirdupois pounds or ounces, except that ready-wound bobbins that are not sold separately shall not be required to be individually marked to show the number of bobbins contained therein and the net meters and yards of thread on each bobbin.

10.10. Packaged Seed. – Packages of seeds intended for planting with net contents of less than 225 g or 8 oz shall be labeled in full accord with this regulation except as follows:

(a) The quantity statement shall appear in the upper 30 % of the principal display panel.

(b) The quantity statement shall be in terms of:

(1) the largest whole SI unit for all packages with weights up to 7 g; and

(2) in grams and ounces for all other packages with weights less than 225 g or 8 oz.
(Amended 1995)

(c) The quantity statement for coated seed, encapsulated seed, pelletized seed, pre-planters, seed tapes, etc., shall be in terms of count.
(Added 1972) (Amended 1975 and 1993)

Section 11. Exemptions [*NOTE 8*, page 83]

NOTE 8: *Section 11. Exemptions include several requirements that refer only to the historic use of inch-pound units or are direct restatements of exemptions contained in federal laws or regulations which do not include SI units. SI equivalents are omitted in most of these requirements because the SI units would not be meaningful or useful.*
(Added 1993) (Amended 1995)

11.1. Random Packages. – A random package bearing a label conspicuously declaring:

(a) the net weight;

(b) unit price; and

(c) the total price

shall be exempt from the SI units, type size, location, and free area requirements of this regulation. In the case of a random package packed at one place for subsequent sale at another, neither the price per unit of weight nor the total selling price need appear on the package, provided the package label includes both such prices at the time it is offered or exposed for sale at retail.

This section shall also apply to uniform weight packages of fresh fruit or vegetables labeled by count, in the same manner and by the same type of equipment as random packages exempted by this section, and cheese and cheese products labeled in the same manner and by the same type of equipment as random packages exempted by this section.
(Amended 1989)

11.1.1. Indirect Sale of Random Packages. – A random package manufactured or produced and offered for indirect sale (e.g., e-commerce, online, phone, fax, catalog, and similar methods) shall be exempt from the labeling requirements of:

(a) unit price

(b) total price

when the following requirements are met:

At the time of the delivery, each package need only bear a statement of net weight, provided that:

(a) the unit price is set forth and established in the initial product offering;

(b) the maximum possible net weight, unit price, and maximum possible price are provided to the customer by order confirmation when the product is ordered; and

(c) when the product is delivered, the customer receives a receipt bearing the following information: identity, declared net weight, unit price, and the total price.

Indirect Sales: For the purpose of Section 11.1.1. Indirect Sale of Random Packages, indirect sales are sales where the customer makes a selection and places an order, but cannot be present when the determination of the

net quantity is made. Examples of such indirect methods include, without limitation, Internet or online sales, sales conducted by telephone or facsimile, and catalog sales.
(Added 2001) (Amended 2002)

11.2. Small Confections. – Individually wrapped pieces of "penny candy" and other confectionery of less than 15 g or ½ oz net weight per individual piece shall be exempt from the labeling requirements of this regulation when the container in which such confectionery is shipped is in conformance with the labeling requirements of this regulation. Similarly, when such confectionery items are sold in bags or boxes, such items shall be exempt from the labeling requirements of this regulation, including the required declaration of net quantity of contents, when the declaration of the bag or box meets the requirements of this regulation.

11.3. Small Packages of Meat or Meat Products. – Individually wrapped and labeled packages of meat or meat products of less than 15 g or ½ oz net weight, which are in a shipping container, need not bear a statement of the net quantity of contents when the statement of the net quantity of contents on the shipping container is in conformance with the labeling requirements of this regulation.
(Added 1987)

11.4. Individual Servings. – Individual serving size packages of foods containing less than 15 g or ½ oz or less than 15 mL or ½ fl oz for use in restaurants, institutions, and passenger carriers, and not intended for sale at retail, shall be exempt from the required declaration of net quantity of contents specified in this regulation.

11.5. Cuts, Plugs, and Twists of Tobacco and Cigars. – When individual cuts, plugs, and twists of tobacco and individual cigars are shipped or delivered in containers that conform to the labeling requirements of this regulation, such individual cuts, plugs, and twists of tobacco and cigars shall be exempt from such labeling requirements.

11.6. Reusable (Returnable) Glass Containers. – Nothing in this Regulation shall be deemed to preclude the continued use of reusable (returnable) glass containers, provided such glass containers ordered after the effective date of this regulation shall conform to all requirements of this regulation.

11.7. Cigarettes and Small Cigars. – Cartons of cigarettes and small cigars, containing ten individual packages of twenty, labeled in accordance with the requirements of this regulation shall be exempt from the requirements set forth in Section 8.1.1. Location, Section 8.2.1. Minimum Height of Numbers and Letters, and Section 10.4. Multi-unit Packages, provided such cartons bear a declaration of the net quantity of commodity in the package.

11.8. Packaged Commodities with Labeling Requirements Specified in Federal Law. – Packages of meat and meat products, poultry products, tobacco and tobacco products, pesticides, and alcoholic beverages shall be exempt from those portions of these regulations specifying location and minimum type size of the net quantity declaration, provided quantity labeling requirements for such products are specified in federal law so as to follow reasonably sound principles of providing consumer information. (See also Section 11.32. SI Units, Exemptions - Consumer Commodities.)

11.9. Fluid Dairy Products, Ice Cream, and Similar Frozen Desserts.

(a) When packaged in ½ liq pt and ½ gal containers, are exempt from the requirements for stating net contents of 8 fl oz and 64 fl oz, which may be expressed as ½ pt and ½ gal, respectively.

(b) When measured by and packaged in measure containers as defined in "Measure Container Code of National Institute of Standards and Technology Handbook 44," are exempt from the requirements of Section 8.1.1. Location that the declaration of net contents be located within the bottom 30 % of the principal display panel.

(c) Milk and milk products when measured by and packaged in glass or plastic containers of ½ pt, 1 pt, 1 qt, ½ gal, and 1 gal capacities are exempt from the placement requirement of Section 8.1.1. Location that the declaration of net contents be located within the bottom 30 % of the principal display panel, provided other required label information is conspicuously displayed on the cap or outside closure and the required net quantity of contents declaration is conspicuously blown, formed, or molded on, or permanently applied to that part of the glass or plastic container that is at or above the shoulder of the container.

(Amended 1993)

11.10. Single Strength and Less than Single-Strength Fruit Juice Beverages, Imitations thereof, and Drinking Water.

(a) When packaged in glass, plastic, or fluid milk type paper containers of 8 fl oz and 64 fl oz capacity, are exempt from the requirements of Section 6.2. Largest Whole Unit to the extent that net contents of 8 fl oz and 64 fl oz (or 2 qt) may be expressed as ½ pt (or half pint) and ½ gal (or half gallon), respectively.

(b) When packaged in glass or plastic containers of ½ pt, 1 pt, 1 qt, ½ gal, and 1 gal capacities, are exempt from the placement requirements of Section 8.1.1. Location that the declaration of net contents be located within the bottom 30 % of the principal display panel, provided other label information is conspicuously displayed on the cap or outside closure and the required net quantity of contents declaration is conspicuously blown, formed, or molded into or permanently applied to that part of the glass or plastic container that is at or above the shoulder of the container.

(Amended 1993)

11.11. Soft Drink Bottles. – Bottles of soft drinks shall be exempt from the placement requirements for the declaration of:

(a) identity when such declaration appears on the bottle closure; and

(b) quantity when such declaration is blown, formed, or molded on or above the shoulder of the container and when all other information required by this regulation appears only on the bottle closure.

11.12. Multi-unit Soft Drink Packages. – Multi-unit packages of soft drinks are exempt from the requirement for a declaration of:

(a) responsibility when such declaration appears on the individual units and is not obscured by the multi-unit packaging or when the outside container bears a statement to the effect that such declaration will be found on the individual units inside; and

(b) identity when such declaration appears on the individual units and is not obscured by the multi-unit packaging.

11.13. Butter. – When packaged in 4 oz, 8 oz, and 1 lb packages with continuous label copy wrapping, butter is exempt from the requirements that the statement of identity (Section 3.1.1. Parallel Identity Declaration: Consumer Package) and the net quantity declaration (Section 8.1.5. Parallel Quantity Declaration) be generally parallel to the base of the package. When packaged in 8 oz and 1 lb units, butter is exempt from the requirement for location (Section 8.1.1. Location) of net quantity declaration.

(Amended 1980 and 1993)

11.14. Eggs. – Cartons containing 12 eggs shall be exempt from the requirement for location (Section 8.1.1. Location) of net quantity declaration. When such cartons are designed to permit division in half, each half shall be exempt from the labeling requirements of this regulation if the undivided carton conforms to all such requirements.

11.15. Flour. – Packages of wheat flour in conventional 2, 5, 10, 25, 50, and 100 lb packages shall be exempt from the requirement in this regulation for location (Section 8.1.1. Location) of the net quantity declaration.

(Amended 1980 and 1993)

11.16. Small Packages. – On a principal display panel of 32 cm^2 (5 in^2) or less, the declaration of quantity need not appear in the bottom 30 % of the principal display panel if that declaration satisfies the other requirements of this regulation.

(Amended 1980)

11.17. Decorative Containers. – The principal display panel of a cosmetic marketed in a "boudoir-type" container, including decorative cosmetic containers of the "cartridge," "pill box," "compact," or "pencil" variety, and those with a capacity of 7.4 mL (¼ oz) or less, may be a tear-away tag or tape affixed to the decorative container and bearing the mandatory label information as required by this regulation.

(Amended 1980)

11.18. Combination and Variety Packages. – Combination and variety packages are exempt from the requirements in this regulation for:

(a) location (see Section 8.1.1. Location);

(b) free area (see Section 8.1.4. Free Area); and

(c) minimum height of numbers and letters (see Section 8.2.1. Minimum Height of Numbers and Letters).

(Amended 1989)

11.19. Margarine. – Margarine in 1 lb rectangular packages, except for packages containing whipped or soft margarine or packages containing more than four sticks, shall be exempt from the requirement in this regulation for location (see Section 8.1.1. Location) of the net quantity declaration.

(Amended 1980 and 1993)

11.20. Corn Flour and Corn Meal. – Corn flour and corn meal packaged in conventional 5, 10, 25, 50, and 100 lb bags shall be exempt from the requirement in this regulation for location (see Section 8.1.1. Location) of the net quantity declaration.

(Amended 1978 and 1980)

11.21. Prescription and Insulin Containing Drugs. – Prescription and insulin containing drugs subject to the provisions of Section 503(b) (1) or 506 of the Federal Food, Drug, and Cosmetic Act shall be exempt from the provisions of this regulation.

11.22. Camera Film, Video Recording Tape, Audio Recording Tape, and Other Image and Audio Recording Media Intended for Retail Sale and Consumer Use. – Image and audio media packaged and labeled for retail sale are exempt from the net quantity statement requirements of this regulation that specify how measurement of commodities should be expressed, provided:

(a) **Unexposed or Unrecorded Media.** – The net quantity of contents of unexposed or unrecorded image and audio media is expressed:

(1) For still film, tape, or other still image media, in terms of the usable or guaranteed number of available still image exposures. The length and width measurements of the individual exposures, expressed in millimeters or inches, are authorized as an optional statement.

Examples:
36 exposures, 36 mm x 24 mm, or
12 exposures, 2¼ in x 2¼ in.

(2) For bulk or movie film, in terms of length (in meters or feet) of film available for exposure.

(3) For all other image and/or audio media, in terms of length of time of electronic media available for recording, together with recording and/or playing speed or other machine settings as necessary. Supplemental information concerning the length of the media [*NOTE 9*, page 87] may be provided.

Supplemental information may be provided on other than the principal display panel.

NOTE 9: *Size, length of media, and format details to ensure interchangeability and other characteristics of audio and imaging media are available in the applicable American National Standards.*

(b) **Exposed, Recorded, or Processed Media.** – The net quantity of contents of exposed or processed film or prerecorded electronic media shall be expressed in terms of the length of time that is of entertainment value.

"Entertainment value" is defined as that portion of a film, tape, or other media that commences with the first frame of sound or picture, whichever comes first after the countdown sequence (if any), and ends with either: (a) the last frame of credits; (b) the last frame of the phrase "The End"; or (c) the end of sound, whichever is last.

(Amended 1990)

11.23. Tint Base Paint. – Tint base paint may be labeled on the principal display panel in terms of a liter, quart, or a gallon, including the addition of colorant selected by the purchaser, provided the system employed ensures that the purchaser always obtains a liter, quart, or a gallon; and further provided, in conjunction with the required quantity statement on the principal display panel, a statement indicating that the tint base paint is not to be sold without the addition of colorant is presented; and further provided the contents of the container, before the addition of colorant, is stated in fluid ounces elsewhere on the label.

Wherever the above conditions cannot be met, containers of tint base paint must be labeled with a statement of the actual net contents prior to the addition of colorant in full accord with all the requirements of this regulation.

(Added 1972) (Amended 1980 and 1993)

11.24. Motor Oil in Cans. – Motor oils when packed in cans bearing the principal display panel on the body of the container are exempt from the requirements of Section 3. Declaration of Identity: Consumer Package to the extent that the Society of Automotive Engineers (SAE) viscosity number is required to appear on the principal display panel, provided the SAE viscosity number appears on the can lid and is expressed in letters and numerals in type size of at least 6 mm or ¼ in.

(Amended 1974, 1980, and 1993)

11.25. Pillows, Cushions, Comforters, Mattress Pads, Sleeping Bags, and Similar Products. – Those products including pillows, cushions, comforters, mattress pads, and sleeping bags, that bear a permanent label as designated by the Association of Bedding and Furniture Law Officials or by the California Bureau of Home Furnishings shall be exempt from the requirements for location (Section 8.1.1. Location), size of letters or numbers (Sections 8.2.1. Minimum Height of Numbers and Letters and 8.2.2. Numbers and Letters: Proportion), free area (Section 8.1.4. Free Area), and the declarations of identity and responsibility (Sections 3.1. Declaration of Identity and 5. Declaration of Responsibility: Consumer and Nonconsumer Packages), provided declarations of identity, quantity, and responsibility are presented on a permanently attached label and satisfy the other requirements of this Regulation, and further provided the information on such permanently attached label be fully observable to the purchaser.

(Added 1973)

11.26. Commodities' Variable Weights and Sizes. – Individual packaged commodities put up in variable weights and sizes for sale intact and intended to be weighed and marked with the correct quantity statement prior to or at the point of retail sale are exempt from the requirements of Section 6. Declaration of Quantity: Consumer Packages while moving in commerce and while held for sale prior to weighing and marking, provided the outside container bears a label declaration of the total net weight.

(Added 1973)

11.27. Packaged Commodities Sold by Count. [*NOTE 10*, page 88] – When a packaged consumer commodity is properly measured in terms of count only, or in terms of count and some other appropriate unit, and the individual units are fully visible to the purchaser, such packages shall be labeled in full accord with this Regulation, except that those containing six or less items need not include a statement of count.

(Added 1973)

NOTE 10: *When the net contents declaration of a package that may enter interstate commerce includes count, federal regulations under the Federal Fair Packaging and Labeling Act provide no exemption from declaring the count unless the count is one (1).*

(Added 1990)

11.28. Textile Packages. – Packages of textiles that are required by Section 6.4.1. Combination Declaration to provide a combination declaration stating the quantity of each individual unit and the count shall be exempt from the requirements in this regulation for:

(a) Location (see Section 8.1.1. Location);

(b) Free area (see Section 8.1.4. Free Area); and

(c) Minimum height of numbers and letters (see Section 8.2.1. Minimum Height of Numbers and Letters).
(Added 1971) (Amended 1989)

11.29. Spot Label. – The declaration of quantity of the contents of a package is exempt from Section 8.1.1. Location requiring the quantity declaration to appear in the bottom 30 % of the principal display panel, as long as the declaration of quantity appears in the lower 30 % of the spot label. In no case may the size of the spot label be used to determine the minimum type size; see Section 8.2. Calculation of Area of Principal Display Panel for Purposes of Type Size for this determination.

(Added 1990)

11.30. Header Strip. – The declaration of quantity of the contents of a package is exempt from Section 8.1.1. Location requiring the quantity declaration to appear in the bottom 30 % of the principal display panel, as long as the declaration of quantity appears in the lower 30 % of the header strip or header label. In no case may the size of the header strip be used to determine the minimum type size; see Section 8.2. Calculation of Area of Principal Display Panel for Purposes of Type Size for this determination.

(Added 1990)

11.31. Decorative Wallcovering Borders. – Decorative wallcovering borders when packaged and labeled for retail sale shall be exempt from the requirements of Sections 6.6.2. One Meter, 1 Square Meter, 1 Kilogram, 1 Liter, 1 Cubic Meter, or More; 6.8.2. One Foot, 1 Square Foot, 1 Pound, 1 Pint, 1 Gallon or More; and 6.9. Bi-dimensional Commodities provided the length and width of the border are presented in terms of the largest whole unit in full accord with the other requirements of the regulation.

(Added 1992) (Amended 1993)

11.32. SI Units, Exemptions - Consumer Commodities. – The requirements for statements of quantity in SI units (except for those in Section 10.10. Packaged Seed and Section 11.22. Camera Film, Video Recording Tape, Audio Recording Tape and Other Image and Audio Recording Media Intended for Retail Sale and Consumer Use) in Section 6. Declaration of Quantity: Consumer Packages shall not apply to:

(a) foods packaged at the retail store level;

(b) random weight packages (see Sections 2.4. Random Package and 11.1. Random Packages);

(c) package labels printed before February 14, 1994;

(d) meat and poultry products subject to the Federal Meat or Poultry Products Inspection Acts;

(e) tobacco or tobacco products;

(f) any beverage subject to the Federal Alcohol Administration Act;

(g) any product subject to the Federal Insecticide, Fungicide, and Rodenticide Act;

(h) drugs and cosmetics subject to the Federal Food, Drug and Cosmetic Act;

(i) nutrition labeling information.

11.33. Inch-Pound Units, Exemptions - Consumer Commodities. – The requirements for statements of quantity in inch-pound units shall not apply to packages that bear appropriate SI units. This exemption does not apply to foods, drugs, or cosmetics or to packages subject to regulation by the FTC, meat and poultry products subject to the Federal Meat or Poultry Products Inspection Acts, and tobacco or tobacco products.

(Added 1999)

Section 12. Variations to be Allowed

12.1. Packaging Variations.

12.1.1. Variations from Declared Net Quantity. – Variations from the declared net weight, measure, or count shall be permitted when caused by unavoidable deviations in weighing, measuring, or counting the contents of individual packages that occur in current good manufacturing practice, but such variations shall not be permitted to such extent that the average of the quantities in the packages of a particular commodity or a lot of the commodity that is kept, offered, or exposed for sale, or sold is below the quantity stated, and no unreasonable shortage in any package shall be permitted even though overages in other packages in the same shipment, delivery, or lot compensate for such shortage. Variations above the declared quantity shall not be unreasonably large.

12.1.2. Variations Resulting from Exposure. – Variations from the declared weight or measure shall be permitted when caused by ordinary and customary exposure to conditions that normally occur in good distribution practice and that unavoidably result in change of weight or measure, but only after the commodity is introduced into intrastate commerce, provided the phrase "introduced into intrastate commerce" as used in this paragraph shall be construed to define the time and the place at which the first sale and delivery of a package is made within the state, the delivery being either:

(a) directly to the purchaser or to his/her agent; or

(b) to a common carrier for shipment to the purchaser,

and this paragraph shall be construed as requiring that so long as a shipment, delivery, or lot of packages of a particular commodity remains in the possession or under the control of the packager or the person who introduces the package into intrastate commerce, exposure variations shall not be permitted.

12.2. Magnitude of Permitted Variations. – The magnitude of package variations of this regulation permitted under Sections 12. Variations to be Allowed, 12.1. Package Variations, 12.1.1. Variations from Declared Net Quantity, and 12.1.2. Variations Resulting from Exposure shall be those expressly set forth in this regulation and variations such as those contained in the procedures and tables of NIST Handbook 133, "Checking the Net Contents of Packaged Goods."
(Amended 1976, 1980, 1984, and 1988)

Section 13. Retail Sale Price Representations

13.1. "Cents off" Representations.

(a) The term "cents off representation" means any printed matter consisting of the words "cents off" or words of similar import (bonus offer, 2 for 1 sale, 1¢ sale, etc.), placed upon any consumer package or placed upon any label affixed or adjacent to such package, stating or representing by implication that it is being offered for sale at a price lower than the ordinary and customary retail sale price.
(Amended 1982)

(b) Except as set forth in Section 13.2. Introductory Offers, the packager or labeler of a consumer commodity shall not have imprinted thereon a "cents off" representation unless:

(1) The commodity has been sold at an ordinary and customary price in the most recent and regular course of business where the "cents off" promotion is made.

(2) The commodity so labeled is sold at a reduction from the ordinary and customary price, which reduction is at least equal to the amount of the "cents off" representation imprinted on the commodity package or label.

(3) Each "cents off" representation imprinted on the package or label is limited to a phrase that reflects that the price marked by the retailer represents the savings in the amount of the "cents off" the retailer's regular price; e.g., "Price Marked is _____ Cents Off the Regular Price," "Price Marked is ___ off the Regular Price of this Package", provided the package or label may in addition bear in the usual pricing spot a form reflecting a space for the regular price, the represented "cents off," and a space for the price to be paid by the consumer.

(4) The commodity at retail presents the regular price, designated as the "regular price", clearly and conspicuously on the package or label of the commodity or on a sign, placard, or shelf marker placed in a position contiguous to the retail display of the "cents off" marked commodity.

i. Not more than three "cents off" promotions of any single size commodity may be initiated in the same trade area within a 12-month period;

ii. At least 30 days must lapse between "cents off" promotions of any particular size packaged or labeled commodity in a specific trade area; and

iii. Any single size commodity so labeled may not be sold in a trade area for a duration in excess of six months within any 12-month period.

(5) Sales of any single size commodity so labeled in a trade area do not exceed in volume 50 % of the total volume of sales of such size commodity in the same trade area during any 12-month period. The 12-month period may be the calendar, fiscal, or market year provided the identical period is applied in this subparagraph and subparagraph (5) of this paragraph. Volume limits may be calculated on the basis of projections for the current year, but shall not exceed 50 % of the sales for the preceding year in the event actual sales are less than the projection for the current year.

(c) No "cents off" promotion shall be made available in any circumstances where it is known or there is reason to know that it will be used as an instrumentality for deception or for frustration of value comparison; for example, where the retailer charges a price that does not fully pass on to the consumers the represented price reduction or where the retailer fails to display the regular price in the display area of the "cents off" marked product.

(d) The sponsor of a "cents off" promotion shall prepare and maintain invoices or other records showing compliance with this section. The invoices or other records required by this section shall be open to inspection and shall be retained for a period of one year subsequent to the end of the year (calendar, fiscal, or market) in which the "cents off" promotion occurs.

(Added 1972)

13.2. Introductory Offers.

(a) The term "introductory offer" means any printed matter consisting of the words "introductory offer" or words of similar import, placed upon a package containing any new commodity or upon any label affixed or adjacent to such new commodity, stating or representing by implication that such new commodity is offered for retail sale at a price lower than the anticipated ordinary and customary retail sale price.

(b) The packager or labeler of a consumer commodity may not have imprinted thereon an introductory offer unless:

(1) The product contained in the package is new, has been changed in a functionally significant and substantial respect, or is being introduced into a trade area for the first time.

(2) Each offer on a package or label is clearly and conspicuously qualified.

(3) No commodity so labeled is sold in a trade area for duration in excess of six months.

(4) At the time of making the introductory offer promotion, the offerer intends in good faith to offer the commodity, alone, at the anticipated ordinary and customary price for a reasonably substantial period of time following the duration of the introductory offer promotion.

(c) The packager or labeler of a consumer commodity shall not have imprinted thereon an introductory offer in the form of a "cents off" representation unless, in addition to the requirements in paragraph (b) of this section:

(1) The package or label clearly and conspicuously and in immediate conjunction with the phrase "Introductory Offer" bears the phrase "__________ cents off the after introductory offer price."

(2) The commodity so labeled is sold at a reduction from the anticipated ordinary customary price, which reduction is at least equal to the amount of the reduction from the after introductory offer price representation on the commodity package or label.

(d) No introductory offer with a "cents off" representation shall be made available in any circumstance where it is known or there is reason to know that it will be used as an instrumentality for deception or for frustration of value comparison; e.g., where the retailer charges a price that does not fully pass on to consumers the represented price reduction.

(e) The sponsor of an introductory offer shall prepare and maintain invoices or other records showing compliance with this section. The invoices or other records required by this section shall be open to inspection and shall be retained for a period of one year subsequent to the period of the introductory offer.

(Added 1972)

13.3. Economy Size.

(a) The term "economy size" means any printed matter consisting of the words "economy size," "economy pack," "budget pack," "bargain size," "value size," or words of similar import placed upon any package containing any consumer commodity or placed upon any label affixed or adjacent to such commodity, stating or representing directly or by implication that a retail sale price advantage is accorded the purchaser thereof by reason of the size of that package or the quantity of its contents.

(b) The packager or labeler of a consumer commodity may not have imprinted thereon an "economy" size representation unless:

(1) At the same time the same brand of the commodity is offered in at least one other packaged size or labeled form.

(2) Only one packaged or labeled form of that brand of commodity labeled with an "economy size" representation is offered.

(3) The commodity labeled with an "economy size" representation is sold at a price per unit of weight, volume, measure, or count that is substantially reduced (i.e., at least 5 %) from the actual price of all other packaged or labeled units of the same brand of that commodity offered simultaneously.

(c) No "economy size" package shall be made available in any circumstances where it is known that it will be used as an instrumentality for deception; e.g., where the retailer charges a price that does not pass on to the consumer the substantial reduction in cost per unit initially granted.

(d) The sponsor of an "economy size" package shall prepare and maintain invoices or other records showing compliance with paragraph (b) of this section. The invoices or other records required by this section shall be open to inspection and shall be retained for one year.

(Added 1972)

Section 14. Revocation of Conflicting Regulations

All provisions of all orders and regulations heretofore issued on this same subject that are contrary to or inconsistent with the provisions of this regulation and specifically __________ are hereby revoked.

Section 15. Effective Date

This regulation shall become effective on __________.

Given under my hand and the seal of my office in the City of __________ on this __________ day of __________.

Signed ____

THIS PAGE LEFT INTENTIONALLY BLANK

UPLR Appendix A: SI/Inch-pound Conversion Factors **

LENGTH							
1 mil (0.001 in)	=	25.4	µm*	1 micrometer	=	0.039 370	mil
1 inch	=	2.54	cm*	1 millimeter	=	0.039 370 1	in
1 foot	=	30.48	cm*	1 centimeter	=	0.393 701	in
1 yard	=	0.914 4	m*	1 meter	=	3.280 84	ft
1 rod	=	5.029 2	m*				
AREA							
1 square inch	=	6.451 6	cm^2*	1 square centimeter	=	0.155 000	in^2
1 square foot	=	929.030	cm^2	1 square decimeter	=	0.107 639	ft^2
1 square yard	=	0.836 127	m^2	1 square meter	=	10.763 9	ft^2
VOLUME or CAPACITY							
1 cubic inch	=	16.387 1	cm^3	1 cubic centimeter	=	0.061 023 74	in^3
1 cubic foot	=	0.028 316 8 28.316 8	m^3 L	1 cubic decimeter	=	0.035 314 7	ft^3
1 cubic yard	=	0.764 555	m^3	1 cubic meter	=	35.314 7 1.307 95	ft^3 yd^3
1 fluid ounce	=	29.573 5	mL	1 milliliter (cm^3)	=	0.033 814	fl oz
1 liquid pint	=	473.177 0.473 177	mL L	1 liter	=	1.056 69 0.264 172	liq qt gal
1 liquid quart	=	946.353 0.946 353	mL L	1 dry pint	=	550.610 5	mL
				1 dry quart	=	1.101 221	L
1 gallon	=	3.785 41	L	1 peck	=	8.809 768	L
1 bushel	=	35.239 1	L	1 gill	=	118.294 1	mL
MASS (weight)							
1 ounce	=	28.349 5	g	1 milligram	=	0.000 035 274 0.015 432 4	oz grain
1 pound	=	453.592 37 0.453 592	g* kg	1 gram	=	0.035 274	oz
1 grain	=	64.798 91	mg	1 kilogram	=	2.204 62	lb
TEMPERATURE							
$t_{°F} = 1.8\ t_{°C} + 32$*				$t_{°C} = \frac{5}{9}(t_{°F} - 32)$*			

* Exactly

** These conversion factors are given to six or more significant digits in the event such accuracy is necessary. To convert to inch-pound units divide the factor rather than multiplying.

(Amended 1998)

THIS PAGE LEFT INTENTIONALLY BLANK

UPLR Appendix B: Converting Inch-pound Units to SI Units for Quantity Declarations on Packages

1. Conversion.

To convert an inch-pound quantity to an SI quantity, multiply the appropriate conversion factor in Table 1 in Appendix A by the inch-pound unit and round according to the following rules.

2. Rounding and Significant Digits.

It is the packager's responsibility to round converted values appropriately and select the appropriate number of significant digits to use in quantity declaration. [These rounding rules are for converting quantity determinations on packages and do not apply to digital scales that automatically round indications to the nearest indicated value.] Conversions, the proper use of significant digits, and rounding must be based on the packer's knowledge of the accuracy of the original measurement that is being converted. For example, if a package is labeled 453.59 g (1 lb), the packer is implying that the package declaration is accurate within ± 0.005 g (or ± 5 mg). For liquid volume measure, a label declaration of 473 mL (16 fl oz) implies that the package declaration is accurate to within ± 0.5 mL (0.01 fl oz). The requirements of 6.13. Rounding apply to all quantity declarations that are derived from converted values:

> **6.13. Rounding.** – In all conversions for the purpose of showing an equivalent SI or inch-pound quantity to a rounded inch-pound or SI quantity, or in calculated values to be declared in the net quantity statement, the number of significant digits retained must be such that accuracy is neither sacrificed nor exaggerated. Conversions, the proper use of significant digits, and rounding must be based on the packer's knowledge of the accuracy of the original measurement that is being converted. In no case shall rounded net contents declarations overstate a quantity; the packer may round converted values down to avoid overstating the net contents.

NOTE: *When as a result of rounding SI or customary inch-pound declarations calculated based on the conversion factors in Appendix A, the resulting declarations are not exact, the largest declaration, whether metric or inch-pound, will be used for enforcement purposes to determine whether a package contains at least the declared amount of the product.*

Do not round conversion factors or any other quantity used or determined in the calculation; only round the final quantity to the number of significant digits needed to maintain the accuracy of the original quantity. Use the rounding rules presented below in Table 1 as guidance to round the final result. In general, quantity declarations on consumer commodities should only be shown to two or three significant digits (for example, 453 g or 85 g). Any final zeros to the right of the decimal point need not be expressed. The inch-pound and SI declarations of quantity must be accurate and equivalent to each other. For example, a package bearing a net weight declaration of 2 lb (32 oz) must also include an SI declaration of 907 g.

Table 1. Rounding Rules		
When The First Digit Dropped is:	**The Last Digit Retained is:**	**Examples**
less than 5	Unchanged	2.44 to 2.4 2.429 to 2.4
more than 5, or 5 followed by at least 1 digit other than 0	Increased by 1	2.46 to 2.5 2.451 to 2.5
5 followed by zeros	Unchanged if Even, or Increased by 1 if Odd	2.450 to 2.4 2.550 to 2.6

(a) When the first digit discarded is less than 5, the last digit retained should not be changed. For example, if the quantity 984.3 is to be declared to 3 significant digits, the number 3 to the right of the decimal point must be discarded since it is less than 5 and the last digit to be retained (the number 4) will remain unchanged. The rounded number will read 984. The same rationale applies to numbers declared to two significant digits (for example 68.4 and 7.34); again, the final digit is dropped and the last digit retained remains unchanged so that the "rounded-off" numbers become 68 and 7.3 respectively.

(b) When the first digit to be discarded is greater than 5, or it is a 5 followed by at least one digit other than zero, the last digit to be retained should be increased by one unit.

Examples:

984.7	becomes	985
984.51	becomes	985
6.86	becomes	6.9
6.88	becomes	6.9

(c) When the first digit to be discarded is exactly 5, followed only by zeros, the final digit to be retained should be rounded up if it is an odd number (1, 3, 5, 7, or 9), but no adjustment should be made if it is an even number (2, 4, 6, or 8).

Examples:

984.50	becomes	984
985.50	becomes	986
68.50	becomes	68
7.450	becomes	7.4
7.550	becomes	7.6

**NOTE: See additional examples in Table 2.*

3. Additional Advice on Rounding and Significant Digits

(a) These rules require the packer to use good judgment in making decisions on how to round and the number of significant digits to use in quantity declarations. Rounding should always be done in one step; for example, if 16.946 47 g has to be rounded to three significant digits, it should be rounded to 16.9 g, not 16.946 5, then to 16.946, then to 16.95 which would then round to 17.0 g (see rounding rules above).

(b) Do not use rounded SI values to calculate quantities. For example, using 1 in = 25.4 mm, rounded to 25 mm, should not be multiplied by 2 to determine the SI equivalent for 2 in. The SI equivalent for 2 in is determined by multiplying 2 in x 25.4 mm = 50.8 mm, then rounding to 51 mm.

(c) If a dimension given as 8 ft is valid to the nearest $^{1}/_{10}$ in, consider it to mean 96.0 in and treat it as having 3 significant digits. The rounded dimension would then be 2.44 m instead of 2.4 m.

(d) Conversions using a multiple digit conversion factor usually give a product with more digits than the original quantity. The final product should contain no more significant digits than are contained in the number with the fewest significant digits used in the conversion. For example, the area of a sheet of paper is determined on a calculator by multiplying 1.25 cm (length) x 1.5 cm (width) = 1.875 cm2. The product given to 4 significant digits on the calculator cannot be any more accurate than two significant digits (the number of significant digits in 1.5 cm), so the area should be declared as 1.9 cm2.

(e) Packagers of consumer commodities should be aware that when a converted value is rounded up, there may be a need to (1) increase the package contents and/or, (2) select a converted value that does not exaggerate the precision of the quantity or overstate the net contents. For example, under the rules above, a net weight declaration of 16 oz (453.592 37 g) may be rounded up to 454 g for three significant digits. Inspections by weights and measures officials are typically conducted using devices with a resolution of 0.5 g or less. If the packer does not address this possibility, some lots of commodities may pass when the inch-pound declaration is tested, but fail when the SI declaration is verified.

Table 2. Examples		
Weight: To convert ounces to grams, multiply ounces by 28.349 5 grams		
Inch-pound	**Calculated SI**	**Rounded SI**
1.0 oz	28.349 5 g	28 g
5.0 oz	141.747 6 g	142 g
10¼ oz	290.582 38 g	291 g*
16.0 oz	453.592 4 g	454 g*
32.0 oz	907.184 g	907 g
48.0 oz	1360.776 g	1.36 kg
5 lb	2.267 962 kg	2.27 kg*
10 lb	4.535 924 kg	4.54 kg*
25 lb	11.339 81 kg	11.3 kg
Liquid Volume: to convert fluid ounces to millimeters, multiply fluid ounces by 29.573 5 milliliters		
Inch-pound	**Calculated SI**	**Rounded SI**
1.0 fl oz	29.573 5 mL	30 mL*
8.0 fl oz	236.588 mL	237 mL*
16.0 fl oz	473.176 mL	473 mL
32.0 fl oz	946.353 mL	946 mL
1 gal	3.785 41 L	3.79 L*
2½ gal	9.463 525 L	9.46 L
5 gal	18.927 05 L	18.9 L
Dry Measure: to convert dry pints to milliliters, multiply dry pints by 550.610 5 milliliters		
Inch-pound	**Calculated SI**	**Rounded SI**
1 dry pt	550.610 5 mL	551 mL*
1 dry qt	1.101 221 L	1.1 L
Length: to convert inches to millimeters, multiply inches by 25.4 millimeters		
Inch-pound	**Calculated SI**	**Rounded SI**
10.5 in	266.7 mm	267 mm* or 26.7 cm*
1 ft	30.48 cm	305 mm* or 30.5 cm*
5 ft	152.4 cm	152 cm or 1.5 m
50 ft	15.240 03 m	15.2 m
100 ft	30.480 06 m	30.5 m*
* See 6.13. Rounding located under UPLR Appendix B		

B. Uniform Regulation for the Method of Sale of Commodities

as adopted by
The National Conference on Weights and Measures*

1. Background

The National Conference on Weights and Measures (NCWM) has long been concerned with the proper units of measurement to be used in the sale of all commodities. This approach has gradually broadened to concerns of standardized package sizes and general identity of particular commodities. Requirements for individual products were at one time made a part of the Weights and Measures Law or were embodied in separate individual Model Regulations. In 1971, this "Model State Method of Sale of Commodities Regulation" was established (renamed in 1983); amendments have been adopted by the Conference almost annually since that time.

Sections with "added 1971" dates refer to those sections that were originally incorporated in the Weights and Measures Law or in individual Model Regulations recommended by the NCWM. Subsequent dates reflect the actual amendment or addition dates.

The 1979 edition included, for the first time, requirements for items packaged in quantities of the International System of Units (SI), the modernized metric system, as well as continuing to present requirements for inch-pound quantities. It should be stressed that nothing in this Regulation requires changing to the SI system of measurement. SI values are given for the guidance of those wishing to adopt new SI quantities of the commodities governed by this Regulation. SI means the International System of Units as established in 1960 by the General Conference on Weights and Measures and interpreted or modified for the United States by the Secretary of Commerce.

This Regulation assimilates all of the actions periodically taken by the Conference with respect to certain food items, non-food items, and general method of sale concepts. Its format is such that it will permit the addition of individual items at the end of appropriate sections as the need arises. Its adoption as a regulation by individual jurisdictions will eliminate the necessity for legislative consideration of changes in the method of sale of particular commodities. Such items will be able to be handled through the normal regulation-making process.

2. Status of Promulgation

The table beginning on page 10 shows the status of adoption of the Uniform Regulation for the Method of Sale of Commodities.

**The National Conference on Weights and Measures (NCWM) is supported by the National Institute of Standards and Technology (NIST) in partial implementation of its statutory responsibility for "cooperation with the states in securing uniformity in weights and measures laws and methods of inspection."*

THIS PAGE INTENTIONALLY LEFT BLANK

Uniform Regulation for the Method of Sale of Commodities

Table of Contents

Section **Page**

THIS PAGE INTENTIONALLY LEFT BLANK

Uniform Regulation for the Method of Sale of Commodities

Preamble

The purpose of this regulation is to require accurate and adequate information about commodities so that purchasers can make price and quantity comparisons.

(Added 1989)

Section 1. Food Products [*NOTE 1*, page 109]

1.1. Berries and Small Fruits

1.1.1. Definitions. – "Small fruits" includes, but is not limited to, cherries, currants, and cherry tomatoes. "Berries" includes all fruit whose names end in the term "berry."

(Added 1991)

NOTE 1: *Packages subject to this Section and the Federal Fair Packaging and Labeling Act shall be labeled in units of the International System of Units (SI) and inch-pound systems of measure effective February 14, 1994, [except for seed (see Section 10.10. Packaged Seed) and camera film and recording tape (see Section 11.22. Camera Film, Video Recording Tape, Audio Recording Tape, and Other Image and Audio Recording Media Intended for Retail Sale and Consumer Use), and as specified in the Uniform Packaging and Labeling Regulation under Section 11.32. SI Units, Exemptions - Consumer Commodities]. SI units may appear first.*

(Added 1982) (Amended 1990 and 1993)

1.1.2. Methods of Sale. – Berries and small fruits shall be offered and exposed for sale and sold by weight [*NOTE 2*, page 109] or by volume. If sold by volume, they must:

(Amended 1991)

(a) be in measure containers that are either open or else covered by uncolored transparent lids or other wrappings that do not obscure the contents, and

(b) have capacities per Section 1.1.2.(b)(1) or Section 1.1.2.(b)(2). When selling berries and small fruits by volume in measure containers, whether or not covered, the measure containers themselves shall not be packages for labeling purposes.

(1) SI Capacities – 250 milliliters, 500 milliliters, or 1 liter.

(Added 1979) (Amended 1985)

(2) Inch-pound Capacities – ½ dry pint, 1 dry pint, or 1 dry quart.

NOTE 2: *When used in this regulation, the term "weight" means "mass." (See paragraphs U. "Mass" and "Weight" in Section I. Introduction of NIST Handbook 130 for an explanation of these terms.)*

1.1.3. Marking Requirements for Shipping Containers. – If two or more measure containers are placed in a shipping package, the crate or package shall show the number of measure containers and the quantity of contents of each.

(Added 1971) (Amended 1979, 1985, 1989, and 1991)

1.2. Bread. – Bread kept, offered, or exposed for sale, whether or not packaged or sliced, shall be sold by weight. The wrappers of bread that is sold and expressly represented at the time of sale as "stale bread" shall not be considered packages for labeling purposes.

(Added 1971) (Amended 1979, 1980, 1985, 1987, 1991, and 1992)

1.3. Butter, Oleomargarine, Margarine, Butter-Like, and/or Margarine-Like Spreads. – Shall be offered and exposed for sale and sold by weight. "Butter-like and/or margarine-like spreads" are those products that meet the Federal Standard of Identity for butter or margarine and oleomargarine, except that they contain less than 80 % fat and may contain other safe and suitable ingredients.

(Added 1971) (Amended 1979, 1985, 1986, and 1994)

1.4. Flour, Corn Meal, and Hominy Grits. – Wheat flour, whole wheat flour, graham flour, self-rising wheat flour, phosphated wheat flour, bromated flour, corn flour, corn meal, and hominy grits, whether enriched or not, shall be packaged, kept, offered, or exposed for sale and sold by weight.

(Amended 1994)

1.5. Meat, Poultry, Fish, and Seafood. [*NOTE 3*, page 110] – Shall be sold by weight, except that whole shellfish in the shell may be sold by weight, measure, and/or count. Shellfish are aquatic animals having a shell, such as mollusks (for example, scallops) or crustaceans (for example, lobster or shrimp).

(Amended 1988)

NOTE 3: *See Section 1.12. Ready-to-Eat Food for additional requirement.*

1.5.1. In Combination with Other Foods. – When meat, poultry, fish, or seafood is combined with some other food element to form a distinctive food product, the quantity representation may be in terms of the total weight of the product or combination, and a quantity representation need not be made for each element provided a statement listing the ingredients in order of their predominance by weight must also appear on the label.

Note: See Interpretations and Guidelines Section 2.2.13. Declaration of Identity: Consumer Package and Labeling Regulation (UPLR).

(Amended 1989)

1.5.2. Clams, Mussels, Oysters, and Other Mollusks.

1.5.2.1. Whole Clams, Oysters, Mussels, or Other Mollusks in the Shell (fresh or frozen). – Shall be sold by weight (including the weight of the shell, but not including the liquid or ice packed with them), dry measure (e.g., bushel), and/or count. In addition, size designations may be provided.

1.5.2.2. Whole Clams, Oysters, Mussels, or Other Mollusks on the Half Shell (fresh, cooked, smoked, or frozen, with or without sauces or spices added). – Shall be sold by weight (excluding the weight of the shell) or by count. Size designations may also be provided.

(Added 1989)

1.5.2.3. Fresh Oysters Removed from the Shell. – Shall be sold by weight, drained weight, or by fluid volume. For oysters sold by weight or by volume, a maximum of 15 % free liquid by weight is permitted.

(Amended 1991)

1.5.2.4. Processed Clams, Mussels, Oysters, or Other Mollusks on the Half Shell (fresh or frozen). – Shall be sold by net weight excluding the weight of the shell. The term "processed" means removing the meat from the shell and chopping it or cutting it or commingling it with other solid foods.

(Amended 1989)

1.5.2.5. Canned (heat-processed) Mussels, Clams, Oysters, or Other Mollusks. – Shall be sold by net weight. A maximum of 41 % free liquid by weight is permitted for canned oysters.

(Added 1986 and 1971) (Amended 1982, 1985, 1986, and 1989)

1.6. Fluid Milk Products. – All fluid milk products, including, but not limited to milk, lowfat milk, skim milk, cultured milks, and cream, shall be sold in terms of fluid volume.

(Amended 1995)

1.7. Other Milk Products. – Cottage cheese, cottage cheese products, and other milk products that are solid, semi-solid, viscous, or a mixture of solid and liquid, as defined in the Pasteurized Milk Ordinance of the U.S. Public Health Service, as amended in 1965, shall be sold in terms of weight.

(Amended 1995)

1.7.1. Factory Packaged Ice Cream and Similar Frozen Products. – Ice cream, ice milk, frozen yogurt, and similar products shall be kept, offered, or exposed for sale or sold in terms of fluid volume.

(Amended 1995)

1.7.2. Pelletized Ice Cream and Similar Pelletized Frozen Desserts. – A semi-solid food product manufactured at very low temperatures using a nitrogen process and consisting of small beads of varying sizes. Bits of inclusions (cookies, candy, etc.) that also vary in size and weight may be mixed with the pellets.

1.7.2.1. Method of Retail Sale. – Packaged pelletized ice cream or similar pelletized frozen desserts shall be kept, offered, or exposed for sale on the basis of net weight.

Note: This method of sale for pelletized ice cream shall be enforceable after April 17, 2010, and after August 2, 2011, for similar pelletized frozen desserts.

(Added 2010) (Amended 2011)

1.8. Pickles. – The declaration of net quantity of contents on pickles and pickle products, including relishes but excluding one or two whole pickles in a transparent wrapping, which may be declared by count, shall be expressed in terms of liquid measure. Sales of pickles from bulk may be by count.

(Added 1971)

1.9. Advertising and Price Computing of Bulk Food Commodities.

1.9.1. Total Price Computing. – The price of food commodities sold from bulk by weight shall be computed in terms of whole units of weight (i.e., grams, kilograms, pounds, ounces, etc.) and not in common or decimal fractions.

1.9.2. Unit Price Advertising. – The price of food commodities sold from bulk by weight shall be advertised or displayed in terms of whole weight units of kilograms or pounds only, not in common or decimal fractions or in ounces. A supplemental declaration is permitted in print no larger than the whole unit price. This supplemental declaration may be expressed in common or decimal fractions or in ounces.

(Added 1976) (Amended 1985, 1987, and 1991)

1.10. Generic Terms for Meat Cuts. – A declaration of identity for meat cuts shall be limited to generic terms, such as those listed in the Uniform Retail Meat Identity Standards.

The following abbreviations may be used:

BAR B Q	Barbecue	POT-RST	Pot Roast
BI	Bone In	RND	Round
BNLS	Boneless	RST	Roast
DBLE	Double	SHLDR	Shoulder
LGE	Large	SQ	Square
N.Y. (NY)	New York	STK	Steak
PK	Pork	TRMD	Trimmed

(Added 1976)

1.11. Sale of Meat by Carcass, Side, or Primal Cut. – The seller of a carcass, side, quarter, or primal cut on a gross or hanging weight basis shall provide to the buyer a written statement giving the following information at the times indicated:
(Amended 1985)

1.11.1. Prior to Delivery.

(a) the name and address of the seller (firm);

(b) the date of the contract;

(c) the name and address of the buyer;

(c) the total net weight (hanging weight) of the carcass, side, or primal cut prior to cutting or processing;

(e) the USDA quality grade and yield grade of the meat to be supplied, if so represented;

(f) the price per pound for each species (not including any inducements) and the total price of the sale order;

(g) the estimated cutting loss on the order in terms of percentage and weight (e.g., 40 %, 72.5 kg [160 lb]);

(h) a list by name and estimated count of each cut to be derived from each primal source;

(i) additional costs, listed separately, for cutting, wrapping, freezing, and finance charges, if any; and

(j) that the buyer may keep the cutting loss.

(Added 1985)

1.11.2. At the Time of Delivery.

(a) the name and address of the buyer and seller;

(b) the date of delivery;

(c) the total net weight of the meat delivered;

(d) a list, by name and count, of each cut derived from each primal cut; and

(e) a separate indication of the quantity of any meat or other commodity(s) received by the purchaser as an inducement in connection with the purchase of the carcass, side, or primal cut.

(Added 1985)

1.11.3. Exemptions. – This subsection shall not apply to the sale of any carcass, side, quarter, or primal cut of meat that individually or collectively has a gross or hanging weight of 22.6 kg (50 lb) or less.
(Added 1985)

1.11.4. Right of Cancellation. – The buyer shall have the right to cancel any carcass, side, quarter, or primal cut meat contract until midnight of the third business day after the day on which the buyer executed the contract or after the day on which the seller provided the buyer with a fully executed copy of the contract, whichever is later.
(Added 1985 and 1977) (Amended 1980 and 1985)

1.12. Ready-to-Eat Food.

1.12.1. Definition - Ready-to-Eat Food. – Restaurant style food offered or exposed for sale, whether in restaurants, supermarkets, or similar food service establishments, that is ready for consumption, though not necessarily on the premises where sold. Ready-to-Eat Food does not include sliced luncheon products, such as meat, poultry, or cheese when sold separately.

NOTE: *The sale of an individual piece of fresh fruit (like an apple, banana, or orange) is allowed by count.*
(Added 2004)

1.12.2. Methods of Sale. – Ready-to-Eat Food sold from bulk or in single servings packed on the premises may be sold by weight, measure, or count (count includes servings).
(Amended 1993)

1.13. Home Food Service Plan Sales.

1.13.1. Definitions.

As used in this section, the following words and phrases shall have the following meanings:

(a) **Home Food Service Plan.** – The offering for sale to a consumer, in the consumer's home, any food item, or food item in combination with any nonfood item and/or services, whether or not a membership fee or similar charge is involved.

(b) **Seller.** – Any person, partnership, corporation, or association, however organized, engaged in the sale of a home food service plan.

(c) **Buyer.** – Both the actual and prospective purchaser, but does not include persons purchasing for resale.

(d) **Contract.** – All of the collective written agreements subscribed by a buyer at the time of sale relating to the purchase of a home food service plan, except promissory notes or other financing agreements.

(e) **Food Item.** – Each edible product sold as part of a home food service plan, including, but not limited to, each constituent part or kind of meat cut from a primal source, each kind of whole poultry or poultry part, seafood products, and other like products.

(f) **Nonfood Item.** – Each inedible product sold as part of a home food service plan, including, but not limited to, paper products, health and beauty products, detergents, cleaners and disinfectants, rolls of wrapping, and like products. The term does not include food items and durable consumer goods such as appliances.

(g) **Unit Price.** – The price of a food or nonfood item sold as part of a home food service plan, computed to the nearest tenth of 1 cent when less than 1 dollar, and to the nearest cent when 1 dollar or more. The unit price, exclusive of any service charge(s), shall be expressed in terms of the price per unit of weight, measure, or count set forth in the "Uniform Unit Pricing Regulation" in the current edition of NIST Handbook 130.

(h) **Service Charge.** – The total price for any additional features, services, and processing associated with the purchase of a home food service plan, whether stated in terms of membership fees or otherwise.

(i) **Primal Source.** – Refers to the following cuts:

(1) for beef, the primal sources are the round, flank, loin, rib, plate, brisket, chuck, and shank;

(2) for veal and lamb or mutton, the primal sources are the leg, flank, loin, rack (rib), and shoulder; and

(3) for pork, the primal sources are the belly, loin, ham, spareribs, shoulder, and jowl.

1.13.2. Contract and Disclosure Requirements.

1.13.2.1. At the Time of Sale:

(a) At the time of sale, the Seller shall provide the Buyer with a single document, referred to in this subsection as the "written agreement," which shall clearly and conspicuously disclose the following:

(1) the name, address, and telephone number of the Seller and the name and address of the Buyer;

(2) the date of the contract;

(3) the price of the food and nonfood items of the home food service plan;

(4) the service charge or the price of any service charges associated with the home food service plan;

(5) the total price of the home food service plan, including the price of the food and nonfood items, and the price of any service charge; and

(6) a statement that the Buyer shall have the right to cancel the home food service plan contract until midnight of the third business day after the date on which the Buyer executed the contract or after the day on which the Seller provided the Buyer with a fully executed copy of the contract, whichever is later, by giving written notice of cancellation to the Seller. Compliance with requirements of federal statutes, rules, or regulations governing form of notice of right of cancellation shall be deemed satisfactory notice of the requirements of this regulation.

(b) In addition to the above disclosures required in the written agreement, the following disclosures are required to be given to the Buyer at the time of sale:

(1) A written list of all food and nonfood items to be sold, which shall include:

i. the identity of each unit and, where applicable, the USDA quality grade of the item, if so graded; the primal source; and the brand or trade name;

ii. the quantity of each item sold;

iii. the estimated serving size by net weight of each piece of meat, poultry, and seafood item offered for sale under the home food service plan, provided, however, that such estimates shall not differ from the actual weight at the time of delivery by more than 5 % and the dollar value of the meat, poultry, and seafood items delivered is equal to or greater than that represented to the Buyer; and

iv. the net weight, measure, or count of all other food and nonfood items offered for sale.

(2) A current unit price list stating in dollars and cents the price per kilogram or pound or other appropriate unit of measure, and the total sale price of each item to be delivered. This price list shall clearly and conspicuously make reference to the fact of whether there are additional costs disclosed in the written agreement relating to any "service charges" associated with the purchase of the home food service plan.

(3) If a membership is sold, a written statement of all terms, conditions, benefits, and privileges applicable to the membership.

(4) If a service charge is included, a written statement specifically identifying the service(s) provided and the price(s) charged for them.

1.13.2.2. At the Time of Delivery:

(a) At the time of delivery, the Seller shall provide a receipt, for signature by the Buyer, disclosing the following information:

(1) the identity of the item and the net quantity of the contents in terms of either weight, measure, or count, as required by applicable law. The net weight of each food item delivered shall be within the limit specified in Section 1.13.2.1.b(i)(iii) Contract and Disclosure Requirements; and

(2) the unit price and total sales price of each food and nonfood item. The unit price shall be the same as that specified on the unit price list given to the Buyer at the time of sale.

1.13.3. Advertisement of Home Food Service Plans. – Any advertisement of a home food service plan which discloses item pricing information in accordance with the provisions of this section shall set forth, in a clear and conspicuous manner, whether there are any service charges or other additional costs associated with the purchase of the home food service plan.
(Added 1992)

Section 2. Non-food Products [*NOTE 1*, page 109]

2.1. Advertising and Price Computing of Bulk Commodities. – The price of bulk commodities or commodities not in package form and sold by weight shall be advertised, displayed, and computed in terms of whole units of weight (i.e., grams, kilograms, pounds, ounces, etc.), and not in common or decimal fractions.
(Added 1989)

2.2. Fence Wire Products. – Rolls of fence wire products shall be sold by:

(a) Gauge of wire.

(b) Height in terms of inches or centimeters, if applicable.

(c) Length in terms of rods, meters, or feet.
(Added 1979)

2.3. Coatings. – Asphalt paints, coatings, and plastics shall be sold in terms of liquid measure.
(Added 1971)

2.4. Fireplace and Stove Wood. – For the purpose of this regulation, this section shall apply to the sale of all wood, natural and processed, for use as fuel or flavoring.
(Amended 1999)

2.4.1. Definitions.

2.4.1.1. Fireplace and Stove Wood. – Any kindling, logs, boards, timbers, or other wood, natural or processed, split or not split, advertised, offered for sale, or sold for use as fuel.
(Amended 1991)

2.4.1.2. Cord. – The amount of wood that is contained in a space of 128 ft^3 when the wood is ranked and well stowed. For the purpose of this regulation, "ranked and well stowed" shall be construed to mean that pieces of wood are placed in a line or row, with individual pieces touching and parallel to each other, and stacked in a compact manner.

2.4.1.3. Representation. – This shall be construed to mean any advertisement, offering, invoice, or the like that pertains to the sale of fireplace or stove wood.

2.4.1.4. Flavoring Chips. – Any kindling, logs, boards, timbers, or other natural or processed, split or unsplit wood that is advertised, offered for sale, or sold for flavoring smoked or barbequed foods.

(Added 1999)

2.4.2. Identity. – A representation may include a declaration of identity that indicates the species group (for example, 50 % hickory, 50 % miscellaneous softwood). Such a representation shall indicate, within 10 % accuracy, the percentages of each group.

2.4.3. Quantity. – Fireplace and stove wood shall be advertised, offered for sale, and sold only by measure, using the term "cord" and fractional parts of a cord or the cubic meter, except that:

(a) **Packaged natural wood.** – Natural wood offered for sale in packaged form in quantities less than 0.45 m^3 ($^1/_8$ cord or 16 ft^3) shall display the quantity in terms of:

(1) liters, to include fractions of liters; or

(2) cubic inches, if less than one cubic foot; or

(3) cubic feet, if one cubic foot or greater, to include fractions of a cubic foot.

(Amended 2010)

(b) **Artificial compressed or processed logs.** – A single fireplace log shall be sold by weight, and packages of such individual logs shall be sold by weight plus count.

(c) **Stove wood pellets or chips.** – Pellets or chips not greater than 15 cm (6 in) in any dimension shall be sold by weight. This requirement does not apply to flavoring chips.

(Amended 1976 and 1991)

(d) **Flavoring chips.** – Flavoring chips offered for sale in packaged form in quantities less than 0.45 m^3 ($^1/_8$ cord or 16 ft^3) shall display the quantity in terms of:

(1) liters, to include fractions of liters; or

(2) cubic inches, if less than one cubic foot; or

(3) cubic feet, if one cubic foot or greater, to include fractions of a cubic foot.

(Added 1998)(Amended 2010)

Note: In determining the appropriate Method of Sale, a clear distinction must be made as to whether the wood is being sold primarily as fuel (some wood is sold as fuel but flavoring is a byproduct) or strictly as a wood flavoring.

(Added 2010)

2.4.4. Prohibition of Terms. – The terms "face cord," "rack," "pile," "truckload," or terms of similar import shall not be used when advertising, offering for sale, or selling wood for use as fuel.

2.4.5. Delivery Ticket or Sales Invoice. – A delivery ticket or sales invoice shall be presented by the seller to the purchaser whenever any non-packaged fireplace or stove wood is sold. The delivery ticket or sales invoice shall contain at least the following information:

(a) the name and address of the vendor;

(b) the name and address of the purchaser;

(c) the date delivered;

(d) the quantity delivered and the quantity upon which the price is based, if this differs from the delivered quantity;

(e) the price of the amount delivered; and

(f) the identity, in the most descriptive terms commercially practicable, including any quality representation made in connection with the sale.

(Added 1975)

2.5. Peat and Peat Moss. – Applies only with respect to organic matter of geological origin, excluding coal and lignite, originating principally from dead vegetative remains through the agency of water in the absence of air and occurring in a bog, swampland, or marsh, and containing an ash content not exceeding 25 % on a dry weight basis [dried in an oven at 105 °C (221 °F) until no further weight loss can be determined].

2.5.1. Declaration of Quantity. – The declaration of quantity of peat and peat moss shall be expressed in weight units or in cubic measure units.

2.5.2. Units.

2.5.2.1. Weight. – Peat and peat moss sold in terms of weight shall be offered and exposed for sale only in kilograms and/or pounds.

2.5.2.2. Cubic Measure. – Peat and peat moss sold in terms of cubic measure shall be offered and exposed for sale only in liters and/or cubic feet. If the commodity is labeled in terms of compressed cubic measurement, the quantity declaration shall represent the quantity in the compressed state.

(Added 1971) (Amended 1975, 1979, 1983, and 1997)

2.6. Prefabricated Utility Buildings. – Shall be offered for retail sale on the basis of usable inside space as follows:

(a) length, measured from inside surface of wall panels at the base;

(b) width, measured from inside surface of wall panels at the base;

(c) height, measured from the base to the top of the shortest wall panel.

Inside dimensions in SI units shall be declared to the nearest 0.01 meter; inside dimensions in inch-pound units shall be declared to the nearest inch.

If total usable inside space is declared in a supplemental declaration, it shall be to the nearest cubic decimeter or cubic foot.

(Added 1975)

2.7. Roofing and Roofing Material. – Shall be sold by the square meter only if sold in SI units, by the square, or by the square foot only if sold in inch-pound units.
(Amended 1979)

2.7.1. Definitions.

2.7.1.1. Square Meter. – The quantity of roofing or roofing material that, when applied according to the directions or instructions of the manufacturer, will cover one square meter exclusive of side laps or side joints.
(Added 1979)

2.7.1.2. Square. – The quantity of roofing or roofing material that, when applied according to directions or instructions of the manufacturer, will cover an area of 100 ft^2 exclusive of side laps or side joints, provided, in the case of roofing or roofing material of corrugated design, the side lap or side joint shall be one full corrugation.

2.7.1.3. Square Foot. – The quantity of roofing or roofing material that, when applied according to the directions or instructions of the manufacturer, will cover 1 ft^2 (144 in^2) exclusive of side laps or side joints.

2.7.2. Declaration of Quantity. – When the declaration of quantity on a package of roofing or roofing material contains the term "square," it shall include, plainly and conspicuously, a numerical definition of the term "square."

Example:
"One square covers 100 ft^2 of roof area."

2.7.2.1. Common Fractions. – The use of the common fraction one-third (⅓) is specifically authorized in the quantity statement of a package of roofing or roofing material when, and only when, used as the common fraction of the "square."

2.7.2.2. Quantity Statement. – The primary declaration if in inch-pound units shall only be in terms of squares or square feet, and if in metric units shall only be in terms of square meters. There is no prohibition against the use of supplementary quantity declarations, such as shingle dimensions, but in no case shall the weight of the material be stated or implied. However, the use of numerical descriptions for rolls of felt roofing material may continue to be used.
(Added 1971) (Amended 1979)

2.8. Sealants. – Caulking compounds, glazing compounds, and putty shall be sold in terms of liquid measure, except that rope caulk shall be sold by weight.
(Added 1971) (Amended 1981)

2.9. Sod and Turf.

2.9.1. Application. – For the purpose of this regulation, this section shall apply to all sod, including turf sod, turf plugs, and turf sprigs.

2.9.2. Definitions.

2.9.2.1. Sod. – Shall mean "turf sod," "turf plugs," or "turf sprigs" of a single kind or variety or a mixture of kinds and varieties.

2.9.2.2. Turf. – The live population of one or more kinds of grasses, legumes, or other plant species used for lawns, recreational use, soil erosion control, or other such purposes.

2.9.2.3. Turf plug. – A small section cut from live turf of those kinds of turf normally vegetatively propagated (such as zoysia grass) that when severed contain sufficient plant material to remain intact.

2.9.2.4. Turf sod. – A strip or section of live turf that when severed contains sufficient plant material to remain intact.

2.9.2.5. Turf sprig. – A live plant, stolon, crown, or section cut from stolonifera plants used as turf.

2.9.3. Quantity. – Sod shall be advertised, offered for sale, and sold by measure or by a combination of count and measure as prescribed by this subsection.

2.9.3.1. Turf sod. – Turf sod shall be advertised for sale and sold in terms of the square meter, square foot, or square yard, as appropriate.

(Amended 1979)

2.9.3.2. Turf plugs. – Turf plugs shall be advertised for sale and sold in terms of count, combined with a statement of the plug diameter.

2.9.3.3. Turf sprigs. – Turf sprigs shall be advertised for sale and sold in terms of the liter or bushel.

(Added 1976) (Amended 1979)

2.10. Softwood Lumber. – Applies to softwood boards, timbers, and dimension lumber that have been surfaced, but shall not apply to rough lumber, to lumber that has been matched, patterned, or shiplapped; or to lumber remanufactured or joined so as to have changed the form or identity, such as individually assembled or packaged millwork items. "Nominal sizes" for inch-pound dimensions are size designations used for convenience in describing approximate, rather than actual, sizes of lumber. "Nominal sizes" were originally derived from the dimensions of rough lumber before surfacing and are always greater than the actual dimensions; thus a dry "2 x 4" is surfaced to actual dimensions of 1½ in x 3½ in (38 mm x 89 mm). The requirements in this section refer to actual sizes of lumber; for nominal sizes (see Table 1. Softwood Lumber Sizes). The nominal sizes used in this section follow Department of Commerce Voluntary Product Standard PS 20-10, "American Softwood Lumber Standard," or latest edition. SI equivalents are included for actual measurements only.

2.10.1. Definitions.

2.10.1.1. Surfaced (dressed) Lumber. – Lumber that has been surfaced by a machine (to attain smoothness of surface and uniformity of size) on one side (S1S), on two sides (S2S), one edge (S1E), two edges (S2E), or a combination of sides and edges (S1S1E, S1S2E, S2S1E, S4S).

2.10.1.2. Boards. – Lumber 38 mm (1½ in) or less in actual thickness and 38 mm (1½ in) or more in actual width. Lumber less than 139 mm (5½ in) in actual width may be classified as strips.

2.10.1.3. Timbers. – Lumber 114 mm (4½ in) or more in smallest dimension. Timbers may be designated as beams, stringers, posts, caps, sills, girders, or purlins.

2.10.1.4. Dimension Lumber. – Lumber from 38 mm (1½ in) to, but not including, 114 mm (4½ in) in actual thickness, and 38 mm (1½ in) or more in actual width. Dimension lumber may be designated as framing, joists, planks, rafters, or studs.

2.10.1.5. Rough Lumber. – Lumber that has not been surfaced, but that has been sawed, edged, and trimmed at least to the extent of showing saw marks, or other primary manufacturing marks in the wood, on the four longitudinal surfaces of each piece for its overall length.

2.10.1.6. Matched Lumber. – Lumber that has been worked with a tongue on one edge of each piece and a groove on the opposite edge to provide a close tongue and groove joint by fitting two pieces together; when end-matched, the tongue and groove are worked in the ends also.

2.10.1.7. Patterned Lumber. – Lumber that is shaped to a pattern or a molded form, in addition to being dressed, matched, or shiplapped, or any combination of these workings.

2.10.1.8. Shiplapped Lumber. – Lumber that has been worked or rabbeted on both edges of each piece to provide a closelapped joint by fitting two pieces together.

2.10.1.9. Grade. – The commercial designation assigned to lumber meeting specifications established by a nationally recognized grade rule writing organization.

2.10.1.10. Species. – The commercial name assigned to a species of trees.

2.10.1.11. Species Group. – The commercial name assigned to two or more individual species having similar characteristics.

2.10.1.12. Representation. – A "representation" shall be construed to mean any advertisement, offering, invoice, or the like that pertains to the sale of lumber.

2.10.1.13. Minimum Dressed Sizes (width and thickness). – The standardized width and thickness at which lumber is dressed when manufactured in accordance with the U.S. Department of Commerce Voluntary Product Standard PS 20-10), "American Softwood Lumber Standard," or latest edition, and regional grading rules conforming to PS 20-10 or latest edition. (See Table 1. Softwood Lumber Sizes.)

2.10.2. Identity. – Representations shall include a declaration of identity that specifies the grade or grades, species or species group, and whether the lumber is unseasoned (green) or dry.

2.10.3. Quantity. – Representations shall be in terms of:

(a) the number of pieces;

(b) the minimum surfaced width and thickness; and

(c) either the length of individual pieces or the lineal footage, except that the use of nominal dimensions shall be allowed as long as a table of minimum surfaced sizes is displayed prominently or the actual dimensions are prominently displayed to the customer and the term "nominal" or "nom" is also used in conjunction with any representation of dimensions.

Table 1. Softwood Lumber Sizes

Minimum standard surfaced sizes at the time of manufacture for both unseasoned (green) and dry lumber as published by the U.S. Department of Commerce in Voluntary Product Standard PS 20-10 or latest edition.

Product Classification (Nominal Size)	Minimum Dressed Sizes**			
	Unseasoned		Dry	
Inches	Inches	Millimeters	Inches	Millimeters
Surfaced Lumber*				
2 x 2	$1\frac{9}{16}$ $1\frac{9}{16}$	40 x 40	1½ x 1½	38 x 38
2 x 2½	$1\frac{9}{16}$ x $2\frac{1}{16}$	40 x 52	1½ x 2	38 x 51
2 x 3	$1\frac{9}{16}$ x $2\frac{9}{16}$	40 x 65	1½ x 2½	38 x 64
2 x 4	$1\frac{9}{16}$ x $3\frac{9}{16}$	40 x 90	1½ x 3½	38 x 89
2 x 6	$1\frac{9}{16}$ x $5\frac{5}{8}$	40 x 143	1½ x 5½	38 x 140
2 x 8	$1\frac{9}{16}$ x 7½	40 x 190	1½ x 7¼	38 x 184
2 x 10	$1\frac{9}{16}$ x 9½	40 x 241	1½ x 9¼	38 x 235
2 x 12	$1\frac{9}{16}$ x 11½	40 x 292	1½ x 11¼	38 x 286
Board Lumber				
1 x 2	$\frac{25}{32}$ x $1\frac{9}{16}$	20 x 40	¾ x 1½	19 x 38
1 x 3	$\frac{25}{32}$ x $2\frac{9}{16}$	20 x 65	¾ x 2½	19 x 64
1 x 4	$\frac{25}{32}$ x $3\frac{9}{16}$	20 x 90	¾ x 3½	19 x 89
1 x 6	$\frac{25}{32}$ x $5\frac{5}{8}$	20 x 143	¾ x 5½	19 x 140
1 x 8	$\frac{25}{32}$ x 7½	20 x 190	¾ x 7¼	19 x 184
1 x 10	$\frac{25}{32}$ x 9½	20 x 241	¾ x 9¼	19 x 235
1 x 12	$\frac{25}{32}$ x 11½	20 x 292	¾ x 11¼	19 x 286

*The dry thicknesses of nominal 3 in and 4 in lumber are 2½ in (64 mm) and 3½ in (89 mm); unseasoned thicknesses are $2\frac{9}{16}$ in (65 mm) and $3\frac{9}{16}$ (90 mm). Widths for these thicknesses are the same as shown above.

**PS 20-10 defines dry lumber as being 19 % or less in moisture content and unseasoned lumber as being over 19 % moisture content. The size of lumber changes approximately 1 % for each 4 % change in moisture content. Lumber stabilizes at approximately 15 % moisture content under normal use conditions.

(Added 1971)

(Added 1971) (Amended 1990 and 1993)

2.11. Carpet. – Anyone who sells carpet shall provide the purchaser with written statements at the time of sale giving the following information:

(a) The name and address of the manufacturer.

(b) The style name and roll number of the carpet.

(c) The generic name of the fiber and the type of backing material.

(d) The amount delivered (exact size shipped).

(e) The price per square meter if sold in SI units, or the price per square foot if sold in inch-pound units, and the total price.

(Added 1977) (Amended 1979 and 1999)

2.12. Hardwood Lumber - Retail Sales. – The requirements of this section apply to retail sales of hardwood lumber, but not to hardwood flooring, molding, or other pre-formed products.

2.12.1. Definitions.

2.12.1.1. Board Foot. – The inch-pound unit of volume measurement for hardwood lumber. A board foot is the volume of a board 1 ft long, 1 ft wide, and 1 in thick or its equivalent (144 in^3 of wood).

2.12.1.2. Surfaced Lumber. – Lumber that has been surfaced for the purpose of attaining smoothness of surface and uniformity of size.

2.12.1.3. Kiln Drying. – A specialized process used to minimize dimensional changes in service. Hardwood lumber used for most products must have moisture removed by placing it in a drying kiln with controlled humidity and heat for a period of time determined by the initial and the final moisture content, the species, and the thickness.

2.12.1.4. Surface Measure. – A rounded area measurement for hardwood lumber. The surface measure shall be determined by multiplying the full width of the piece in inches and fractions by the standard length (see Section 2.12.1.7. Standard Lengths) in feet, dividing by 12, and rounding up or down to the nearest whole square foot. (Fractions less than or equal to one-half square foot are rounded down and those greater than one-half square foot are rounded up.)

2.12.1.5. Species. – The commercial name assigned to a species of trees.

2.12.1.6. Species Group. – The commercial name assigned to two or more individual species having similar characteristics.

2.12.1.7. Standard Lengths. – 4, 5, 6, 7, 8, 9, 10, 11, 12, 13, 14, 15, or 16 feet. Fractional lengths are rounded down to the next lower standard length (for example, if a board is 6 ft 8 in long, its length is rounded down to 6 ft).

2.12.1.8. Stock Widths. – Special items manufactured to predetermined widths, normally for retail sale.

2.12.2. Identity. – Representations shall include a declaration of identity that specifies the species or species group.

2.12.3. Surfaced (S4S) Lumber Manufactured to Stock Widths.

2.12.3.1. Quantity. – Representations shall be in terms of one of the following:

(a) by linear measure when surfaced width and thickness are stated; or

(b) by count when length and surfaced width and thickness are stated; or

(c) by surface measure (square feet) when a thickness is stated.

2.12.3.2. Representations. – The use of nominal dimensions shall be allowed if the table of Minimum Surfaced Sizes for Kiln Dried Hardwood Lumber or the actual dimensions are prominently displayed to the customer, and the term "nominal" or "nom" is used in conjunction with any representation of nominal dimensions.

2.12.3.3 Minimum surfaced sizes for Kiln Dried Lumber (width and thickness). – Table 2. Minimum Surfaced Sizes for Kiln Dried Hardwood Lumber shows the minimum sizes for the stock widths listed. This table includes dimensions for thicknesses of 1 in and 2 in thick stock lumber. Hardwood lumber is also manufactured in thicknesses of 1¼ in (1 in surfaced) and 1½ in (1^3/16 in surfaced). For other thicknesses, use the nominal and minimum widths from the table. For example: a board with the nominal dimensions of 1¼ in x 4 in would have minimum thickness of 1 in and minimum width of 3½ in.

Table 2. Minimum Surfaced Sizes for Kiln Dried Hardwood Lumber		
SI Units for Thickness and Width	**Thickness and Width in Inches**	
Minimum Sizes in Millimeters	**Nominal Sizes**	**Minimum Sizes**
38 x 89	2 x 4	1½ x 3½
38 x 140	2 x 6	1½ x 5½
38 x 184	2 x 8	1½ x 7¼
38 x 235	2 x 10	1½ x 9¼
38 x 286	2 x 12	1½ x 11¼
19 x 19	1 x 1	¾ x ¾
19 x 38	1 x 2	¾ x 1½
19 x 63	1 x 3	¾ x 2½
19 x 89	1 x 4	¾ x 3½
19 x 140	1 x 6	¾ x 5½
19 x 184	1 x 8	¾ x 7¼
19 x 235	1 x 10	¾ x 9¼
19 x 286	1 x 12	¾ x 11¼
The dry thickness of nominal 1½ in lumber is 1^3/16 in. The dry thickness of nominal 1¼ in lumber is 1 in. Sizes are shown in inches and millimeters. Minimum sizes in millimeters are calculated by multiplying the size in inches by 25.4 and rounding to the nearest millimeter. The rule for rounding is: round up for numbers greater than 0.50 mm and round down for numbers less than or equal to 0.50 mm. In case of a dispute on size measurements, the inch measurement takes precedence. Nominal and minimum widths for these thicknesses are shown above. The SI equivalents for 1 in and 1^3/16 in lumber are 25.4 mm and 30.1 mm, respectively.		

2.12.4. Random Width Lumber.

2.12.4.1. Sales of Random Width Hardwood Lumber. – Sales of random width hardware lumber measured after kiln drying shall be quoted, invoiced, and delivered on the basis of net board footage with no addition of footage for kiln drying shrinkage or surfacing. Sales of hardwood lumber measured and sold prior to kiln drying or surfacing shall be quoted, invoiced, and delivered on the basis of net board footage before kiln drying or surfacing. If the lumber is to be kiln dried or surfaced at the request of the purchaser, the kiln drying or surfacing charge shall be clearly shown and identified on the quotation and invoice.

(Amended 1993)

2.13. Polyethylene Products.

2.13.1. Consumer and Non-consumer Products. – Offered and exposed for sale shall be sold in the terms given in Section 2.13.1.1. Sheeting and film.

2.13.1.1. Sheeting and Film.

Consumer products shall include quantity statements in both SI and inch-pound units.

Consumer products:

(a) length and width (in SI and inch-pound units)

(b) area (in square meters and square feet)

(c) thickness (in micrometers and mils [*NOTE 4*, page 124])

(d) weight (in SI and inch-pound units)

Non-Consumer Products:

(a) length and width (in SI or inch-pound units)

(b) area (in square meters or square feet)

(c) thickness (in micrometers or mils [*NOTE 4*, page 124])

(d) weight (in SI or inch-pound units)
(Added 1982) (Amended 1979, 1993, and 1998)

NOTE 4: *1 mil = 0.001 in = 25.4 micrometers (µm). 1 micrometer = 0.000 039 37 in.*
(Amended 1993)

2.13.2. Consumer Products. – at retail shall be sold in the terms given in Section 2.13.2.1. Food wrap, Section 2.13.2.2. Lawn and trash bags, and Section 2.13.2.3. Food and sandwich bags.

2.13.2.1. Food Wrap.

(a) length and width

(b) area in square meters and square feet
(Amended 1979)

2.13.2.2. Lawn and Trash Bags.

(a) count

(b) dimensions

(c) thickness in micrometers and mils
(Amended 1993)

(d) capacity [NOTE 5, page 125]

2.13.2.3. Food and Sandwich Bags.

The capacity statement does not apply to fold-over sandwich bags.

(a) count

(b) dimensions

(c) capacity [NOTE 5, page 125]

NOTE 5: *See Section 10.8.2. Capacity of the Uniform Packaging and Labeling Regulation.*

2.13.3. Non-consumer Products. – Shall be offered and exposed for sale in the terms given in Section 2.13.3.1. Bags. (Package shall be labeled in SI or inch-pound units and may include both units.)
(Amended 1998)

2.13.3.1. Bags.

(a) count

(b) dimensions

(c) thickness in micrometers or mils

(d) weight

(e) capacity [NOTE 5, page 125]

2.13.4. Declaration of Weight. – The labeled statement of weight for polyethylene sheeting and film products under Sections 2.13.1.1. Sheeting and film, and 2.13.3.1. Bags, shall be equal to or greater than the weight calculated by using the formula below. The final value shall be calculated to four digits, and declared to three digits, dropping the final digit as calculated (for example, if the calculated value is 2.078 lb, then the declared net weight shall be 2.07 lb).

For SI dimensions:

M = T x A x D/1000, where:

M = net mass in kilograms
T = nominal thickness in centimeters
A = nominal length in centimeters times nominal width [NOTE 6, page 126] in centimeters
D = minimum density in grams per cubic centimeter as defined by the latest version of ASTM Standard D1505, "Standard Test Method for Density of Plastics by the Density-Gradient

Technique" and the latest version of ASTM Standard D883, "Standards Terminology Relating to Plastics."

For the purpose of this regulation, the minimum density (D) for linear low density polyethylene plastics (LLDPE) shall be 0.92 g/cm^3 (when D is not known).

For the purpose of this regulation, the minimum density (D) for linear medium density polyethylene plastics (LMDPE) shall be 0.93 g/cm^3 (when D is not known).

For the purpose of this regulation, the minimum density (D) for high density polyethylene plastics (HDPE) shall be 0.94 g/cm^3 (when D is not known).

For inch-pound dimensions:

W = T x A x 0.03613 x D, where:

W = net weight in pounds
T = nominal thickness in inches;
A = nominal length in inches times nominal width [*NOTE 6*, page 126] in inches
D = minimum density in grams per cubic centimeter as defined by the latest version of ASTM Standard D1505, "Standard Test Method for Density of Plastics by the Density-Gradient Technique" and the latest version of ASTM Standard D883, "Standards Terminology Relating to Plastics."

0.03613 is a factor for converting g/cm^3 to lb/in^3

For the purpose of this regulation, the minimum density (D) for linear low density polyethylene plastics (LLDPE) shall be 0.92 g/cm^3 (when D is not known).

For the purpose of this regulation, the minimum density (D) for linear medium density polyethylene plastics (LMDPE) shall be 0.93 g/cm^3 (when D is not known).

For the purpose of this regulation, the minimum density (D) for high density polyethylene plastics (HDPE) shall be 0.94 g/cm^3 (when D is not known).
(Added 1977) (Amended 1980, 1982, 1987, 1989, 1990, 1993, and 2012)

NOTE 6: *The nominal width for bags in this calculation is twice the labeled width.*

2.14. Insulation.

2.14.1. Packaged Loose-Fill Insulation Except Cellulose. – The label shall declare:

(a) the type of insulation and the net weight with no qualifying statement; and

(b) the minimum thickness, maximum net coverage area, and minimum weight per square foot at R values of 11, 19, and 22. This information shall also be supplied for any additional R values listed.
(Amended 1990)

2.14.2. Packaged Loose-Fill Cellulose Insulation. – The label shall declare:

(a) the type of insulation and the net weight with no qualifying statement; and

(b) the minimum thickness, maximum net coverage area, number of bags per 1000 ft^2, and minimum weight per square foot at R values of 13, 19, 24, 32, and 40. This information shall also be supplied for any additional R values listed.

(Amended 1990)

2.14.3. Batt and Blanket Insulation. – The principal display panel of packaged batt or blanket insulation shall declare the square feet of insulation in the package and the length and width of the batt or blanket. In addition, R value and thickness shall be declared on the package.

2.14.4. Installed Insulation. – Installed insulation must be accompanied by a contract or receipt. For all insulation except loose-fill and aluminum foil, the receipt must show the coverage area, thickness, and R value of the insulation installed. For loose-fill, the receipt must show the coverage area, thickness, and R value of the insulation, plus the number of bags used. For aluminum foil, the receipt must show the number and thickness of the air spaces, the direction of heat flow, and R value. The receipt must be dated and signed by the installer.

Example: This is to certify that the insulation has been installed in conformance with the requirements indicated by the manufacturer to provide a value of R 19 using 31.5 bags of insulation to cover 1500 ft^2 area. Signed and dated.

(Added 1979) (Amended 1983)

2.15. Solid Fuel Products. – Anthracite, semi anthracite, bituminous, semi-bituminous or lignite coal, and any other natural, manufactured, or patented fuel, not in liquid or gaseous form, except fireplace and stove wood, shall be offered, exposed for sale, or sold by net weight when in package form.

(Added 1979)

2.16. Compressed or Liquefied Gases in Refillable Cylinders.

2.16.1. Application. – This section does not apply to disposable cylinders of compressed or liquefied gases.

2.16.2. Net Contents. – The net contents shall be expressed in terms of cubic meters or cubic feet, kilograms, or pounds and ounces. See Section 2.21. Liquefied Petroleum Gas for permitted expressions of net contents for liquefied petroleum gas. A standard cubic foot of gas is defined as a cubic foot at a temperature of 21 °C (70 °F) and a pressure of 101.35 kilopascals (14.696 psia), except for liquefied petroleum gas as stated in Section 2.21.

2.16.3. Cylinder Labeling. – Whenever cylinders are used for the sale of compressed or liquefied gases by weight, or are filled by weight and converted to volume, the following shall apply:

2.16.3.1. Tare weights.

(a) **Stamped or Stenciled Tare Weight.** – For safety purposes, the tare weight shall be legibly and permanently stamped or stenciled on the cylinder. All tare weight values shall be preceded by the letters "TW" or the words "tare weight." The tare weight shall include the weight of the cylinder (including paint), valve, and other permanent attachments. The weight of a protective cap shall not be included in tare or gross weights. The Code of Federal Regulations Title 49, Section 178.50-22 requires the maker of cylinders to retain test reports verifying the cylinder tare weight accuracy to a tolerance of 1 %.

(b) **Tare Weight for Purposes of Determining the Net Contents.** – The tare weight used in the determination of the final net contents may be either:

(1) the stamped or stenciled tare weight; or

(2) the actual tare determined at the time of filling the cylinder. If the actual tare is determined at the time of filling the cylinder, it must be legibly marked on the cylinder or on a tag attached to the cylinder at the time of filling.

(c) **Allowable difference.** – If the stamped or stenciled tare is used to determine the net contents of the cylinder, the allowable difference between the actual tare weight and the stamped (or stenciled) tare weight, or the tare weight on a tag attached to the cylinder for a new or used cylinder, shall be:

(1) ½ % for tare weights of 9 kg (20 lb) or less; or

(2) ¼ % for tare weights of more than 9 kg (20 lb).

(d) **Average requirement.** – When used to determine the net contents of cylinders, the stamped or stenciled tare weights of cylinders at a single place of business found to be in error predominantly in a direction favorable to the seller and near the allowable difference limit shall be considered to be not in conformance with these requirements.

2.16.3.2. Acetylene Gas Cylinder Tare Weights. – Acetone in the cylinder shall be included as part of the tare weight.

2.16.3.3. Acetylene Gas Cylinder Volumes. – The volumes of acetylene shall be determined from the product weight using approved tables such as those published in NIST Handbook 133 or those developed using 70 °F (21 °C) and 14.7 ft^3 (101.35 kPa) per pound at 1 atmosphere as conversion factors.

2.16.3.4. Compressed Gases such as Oxygen, Argon, Nitrogen, Helium, and Hydrogen. – The volumes of compressed gases such as oxygen, argon, nitrogen, helium, or hydrogen shall be determined using the tables and procedures given in NIST Technical Note 1079, Tables of Industrial Gas Container Contents and Density for Oxygen, Argon, Nitrogen, Helium, and Hydrogen and supplemented by additional procedures and tables in NIST Handbook 133.

(Added 1981) (Amended 1990)

2.17. Precious Metals.

2.17.1. Definition.

2.17.1.1. Precious Metals. – Gold, silver, platinum, or any item composed partly or completely of these metals or their alloys and in which the market value of the metal in the item is principally the gold, silver, or platinum component.

2.17.2. Quantity. – The unit of measure and the method of sale of precious metals, if the price is based in part or wholly on a weight determination, shall be either troy weight or SI units. When the measurement or method of sale is expressed in SI units of mass, a conversion chart to troy units shall be prominently displayed so as to facilitate price comparison. The conversion chart shall also display a table of troy weights indicating grains, pennyweights, and troy ounces.

(Added 1982)

2.18. Mulch.

2.18.1. Definition.

2.18.1.1. Mulch. – Any product or material except peat or peat moss (see Section 2.5. Peat and Peat Moss) that is advertised, offered for sale, or sold for primary use as a horticultural, aboveground dressing, for decoration, moisture control, weed control, erosion control, temperature control, or other similar purposes.

(Added 1987)

2.18.2. Quantity. – All mulch shall be sold, offered, or exposed for sale in terms of volume measure in SI units in terms of the cubic meter or liter or in inch-pound units in terms of the cubic yard or cubic foot.

(Added 1983) (Amended 1987)

2.19. Kerosene (Kerosine). – All kerosene kept, offered, exposed for sale, or sold shall be identified as such and will include, with the word kerosene, an indication of its compliance with the latest version of the standard specification ASTM Standard D3699, "Standard Specification for Kerosine."

Example:
1K Kerosene; Kerosene - 2K.

(Added 1983)

2.19.1. Retail Sale from Bulk. – All kerosene kept, offered, or exposed for sale and sold from bulk at retail shall be in terms of the gallon or liter.

(Added 2012)

2.20. Gasoline-Oxygenate Blends.

2.20.1. Method of Retail Sale. – Type of Oxygenate must be Disclosed. – All automotive gasoline or automotive gasoline-oxygenate blends kept, offered, or exposed for sale, or sold at retail containing at least 1.5 mass percent oxygen shall be identified as "with" or "containing" (or similar wording) the predominant oxygenate in the engine fuel. For example, the label may read "contains ethanol" or "with MTBE." The oxygenate contributing the largest mass percent oxygen to the blend shall be considered the predominant oxygenate. Where mixtures of only ethers are present, the retailer may post the predominant oxygenate followed by the phrase "or other ethers" or alternatively post the phrase "contains MTBE or other ethers." In addition, gasoline-methanol blend fuels containing more than 0.15 mass percent oxygen from methanol shall be identified as "with" or "containing" methanol. This information shall be posted on the upper 50 % of the dispenser front panel in a position clear and conspicuous from the driver's position in a type at least 12.7 mm (½ in) in height, 1.5 mm ($^{1}/_{16}$ in) stroke (width of type).

(Amended 1996)

2.20.2. Documentation for Dispenser Labeling Purposes. – At the time of delivery of the fuel, the retailer shall be provided, on an invoice, bill of lading, shipping paper, or other documentation a declaration of the predominant oxygenate or combination of oxygenates present in concentrations sufficient to yield an oxygen content of at least 1.5 mass percent in the fuel. Where mixtures of only ethers are present, the fuel supplier may identify either the predominant oxygenate in the fuel (i.e., the oxygenate contributing the largest mass percent oxygen) or, alternatively, use the phrase "contains MTBE or other ethers." In addition, any gasoline containing more than 0.15 mass percent oxygen from methanol shall be identified as "with" or "containing" methanol. This documentation is only for dispenser labeling purposes; it is the responsibility of any potential blender to determine the total oxygen content of the engine fuel before blending.

(Added 1984) (Amended 1985, 1986, 1991, and 1996)

2.21. Liquefied Petroleum Gas. – All liquefied petroleum gas, including, but not limited to propane, butane, and mixtures thereof, shall be kept, offered, exposed for sale, or sold by the pound, metered cubic foot [*NOTE 7*, page 127] of vapor (defined as 1 ft^3 at 60 °F [15.6 °C]), or the gallon (defined as 231 in^3 at 60 °F [15.6 °C]). All metered sales by the gallon, except those using meters with a maximum rated capacity of 20 gal/min or less, shall be accomplished by use of a meter and device that automatically compensates for temperature.

(Added 1986)

NOTE 7: *Sources: American National Standards Institute, Inc., "American National Standard for Gas Displacement Meters (500 Cubic Feet per Hour Capacity and Under)," First edition, 1974, and NIST Handbook 44, "Specifications, Tolerances, and Other Technical Requirements for Weighing and Measuring Devices."*

2.22. Liquid Oxygen Used for Respiration.

(a) If sold by weight, liquid oxygen must be weighed on an appropriate, sealed commercial scale. A pressure or other type of gauge may not be used to determine weight.

(b) A delivery ticket or sales invoice shall be provided and shall contain at least the following information:

(1) date delivered;

(2) name and address of vendor;

(3) name and address of the purchaser;

(4) if sold by weight:

i. weight of cylinder before filling;

ii. weight of cylinder after filling; and

iii. the net weight of liquid oxygen delivered;

(5) if sold by measure:

i. measurement and any computation used to arrive at the net quantity of liquid oxygen delivered;

(6) the unit price;

(7) the total computed price; and

(8) weigher's or measurer's signature.
(Added 1989)

2.23. Animal Bedding. – Packaged animal bedding of all kinds, except for baled straw, shall be sold by volume, that is, by the cubic meter, liter, or milliliter and by the cubic yard, cubic foot, or cubic inch. If the commodity is packaged in a compressed state, the quantity declaration shall include both the quantity in the compressed state and the usable quantity that can be recovered. Compressed animal bedding packages shall not include pre-compression volume statements.

Example:
250 mL expands to 500 mL (500 in^3 expands to 1000 in^3).

(Added 1990) (Amended 2012)

2.23.1. Exemption - Non-Consumer Packages Sold to Laboratory Animal Research Industry. – Packaged animal bedding consisting of granular corncobs and other dry (8 % or less moisture), pelleted, and/or non-compressible bedding materials that are sold to commercial (non-retail) end users in the laboratory animal research industry (government, medical, university, preclinical, pharmaceutical, research, biotech, and research institutions) may be sold on the basis of weight.
(Added 2010)

2.24. Wiping Cloths. – Wiping cloths shall be sold by net weight or by count plus size of wiping cloths. When sold by count plus size, and the wiping cloths are of assorted sizes, the term "irregular dimensions" and the minimum size of such cloths must be declared. The gross weight may not be printed on any package, either consumer or non-consumer.

(Added 1991)

2.25. Baler Twine. – Baler twine shall be sold on the basis of length in meters or feet, and net mass or weight by kilograms or pounds.
(Added 1992)

2.26. Potpourri. – Potpourri shall be sold as follows:

(a) Potpourri packaged in advance of sale shall be sold by weight, except when sold in a decorative container or sachet, which may be sold by count.

(b) Potpourri sold from bulk shall be sold by weight or by dry volume.
(Added 1992)

2.27. Retail Sales of Natural Gas Sold as a Vehicle Fuel.

2.27.1. Definitions.

2.27.1.1. Natural Gas. – A gaseous fuel composed primarily of methane that is suitable for compression and dispensing into a fuel storage container(s) for use as an engine fuel.

2.27.1.2. Gasoline Liter Equivalent (GLE). – Gasoline liter equivalent (GLE) means 0.678 kg of natural gas.

2.27.1.3. Gasoline Gallon Equivalent (GGE). – Gasoline gallon equivalent (GGE) means 2.567 kg (5.660 lb) of natural gas.

2.27.2. Method of Retail Sale and Dispenser Labeling.

2.27.2.1. Method of Retail Sale. – All natural gas kept, offered, or exposed for sale and sold at retail as a vehicle fuel shall be in terms of the gasoline liter equivalent (GLE) or gasoline gallon equivalent (GGE).

2.27.2.2. Dispenser Labeling. – All retail natural gas dispensers shall be labeled with the conversion factor in terms of kilograms or pounds. The label shall be permanently and conspicuously displayed on the face of the dispenser and shall have either the statement "1 Gasoline Liter Equivalent (GLE) is equal to 0.678 kg of Natural Gas" or "1 Gasoline Gallon Equivalent (GGE) is equal to 5.660 lb of Natural Gas" consistent with the method of sale used.

2.28. Communication Paper.

2.28.1. Definitions.

2.28.1.1. Communication Paper. – Packaged bond, mimeo, spirit duplicator, xerographic, and other papers, including cut-sized office paper and computer paper.

2.28.1.2. Basis Weight. – As used in this regulation for labeling means the grade, category, or identity of the paper determined according to the latest version of ASTM Standard Method D646, "Grammage of Paper and Paperboard." Basis weight is used as a standard of identity and is not considered a net weight declaration.

2.28.2. Method of Retail Sale and Labeling.

2.28.2.1. Method of Retail Sale. – All packaged communication paper kept, offered, or exposed for sale and sold at retail shall be sold in terms of sheet length and width and count.

2.28.2.2. Labeling. – Communication paper in package form shall bear a label that includes:

(a) a declaration of quantity, in terms of sheet length and width and count, in the lower 30 % of the principal display panel.

(b) a declaration of identity including the basis weight, and may include such other information as grain direction, color, brightness, printed lines, and hole punch information. Due to the variation in basis weight in manufacturing and analysis, the basis weight declared on the label shall correspond to the basis weight declared by the original manufacturer.

(Added 1994)

2.29. Sand, Rock, Gravel, Stone, Paving Stone, and Similar Materials, when Sold in Bulk. – All sand, rock, gravel, stone, paving stone, and similar materials kept, offered, or exposed for sale in bulk must be sold as follows:

(a) Top soil, fill dirt, aggregate or chipped rock, sand (including concrete and mortar sand), decomposed granite, landscape type rock, and cinders must be sold by the cubic meter or cubic yard or by weight.

(b) Flagstone must be sold by weight.

(c) Dimensional cut stone must be sold by square meter, square foot, or weight.

(d) This requirement does not apply to single stones with engraving such as gravestones, natural or manmade artwork, landscape boulders, and pre-cast uniform size blocks.

(Added 2000)

2.30. E85 Fuel Ethanol.

2.30.1. How to Identify Fuel Ethanol. – Fuel ethanol shall be identified as E85.

2.30.2. Labeling Requirements.

(a) Fuel ethanol shall be labeled with its automotive fuel rating in accordance with 16 Code of Federal Regulations Part 306.

(b) A label shall be posted which states "For Use in Flexible Fuel Vehicles (FFV) Only." This information shall be clearly and conspicuously posted on the upper 50 % of the dispenser front panel in a type at least 12.7 mm (½ in) in height, 1.5 mm (1/16 in) stroke (width of type). A label shall be posted which states, "Consult Vehicle Manufacturer Fuel Recommendations," and shall not be less than 6 mm (¼ in) in height by 0.8 mm (1/32 in) stroke; block style letters and the color shall be in definite contrast to the background color to which it is applied.

(Added 2007)

2.31. Biodiesel and Biodiesel Blends.

2.31.1. Identification of Product. – Biodiesel shall be identified by the term "Biodiesel" with the designation "B100." Biodiesel Blends shall be identified by the term "Biodiesel Blend."

2.31.2. Labeling of Retail Dispensers.

2.31.2.1. Labeling of Grade Required. – Biodiesel shall be identified by the grades S15 or S500. biodiesel blends shall be identified by the grades No. 1-D, No. 2-D, or No. 4-D.

2.31.2.2. EPA Labeling Requirements also Apply. – Retailers and wholesale purchaser-consumers of biodiesel blends shall comply with EPA pump labeling requirements for sulfur under 40 CFR § 80.570.

2.31.2.3. Automotive Fuel Rating. – Biodiesel and biodiesel blends shall be labeled with its automotive fuel rating in accordance with 16 CFR Part 306.

2.31.2.4. Biodiesel Blends. – When biodiesel blends greater than 20 % by volume are offered by sale, each side of the dispenser where fuel can be delivered shall have a label conspicuously placed that states "Consult Vehicle Manufacturer Fuel Recommendations." The lettering of this legend shall not be less than 6 mm (¼ in) in height by 0.8 mm (1/32 in) stroke; block style letters and the color shall be in definite contrast to the background color to which it is applied.

2.31.3. Documentation for Dispenser Labeling Purposes. – The retailer shall be provided, at the time of delivery of the fuel, a declaration of the volume percent biodiesel on an invoice, bill of lading, shipping paper, or other document. This documentation is for dispenser labeling purposes only; it is the responsibility of any potential blender to determine the amount of biodiesel in the diesel fuel prior to blending.

2.31.4. Exemption. – Biodiesel blends that contain less than or equal to 5 % biodiesel by volume are exempt from the requirements of Sections 2.31.1. Identification of Product, 2.31.2. Labeling of Retail Dispensers, and 2.31.3. Documentation for Dispenser Labeling Purposes when it is sold as diesel fuel.

(Added 2008)

2.32. Retail Sales of Hydrogen Fuel (H).

2.32.1. Definitions for Hydrogen Fuel. – A fuel composed of molecular hydrogen intended for consumption in a surface vehicle or electricity production device with an internal combustion engine or fuel cell.

(Amended 2012)

2.32.2. Method of Retail Sale and Dispenser Labeling. – All hydrogen fuel kept, offered, or exposed for sale and sold at retail shall be in mass units in terms of the kilogram. The symbol for hydrogen vehicle fuel shall be the capital letter "H" (the word Hydrogen may also be used).

2.32.3. Retail Dispenser Labeling.

(a) A computing dispenser must display the unit price in whole cents on the basis of price per kilogram.

(b) The service pressure(s) of the dispenser must be conspicuously shown on the user interface in bar or the SI unit of pascal (Pa) (e.g., MPa).

(c) The product identity must be shown in a conspicuous location on the dispenser.

(d) National Fire Protection Association (NFPA) labeling requirements also apply.

(e) Hydrogen shall be labeled in accordance with 16 CFR 309 – FTC Labeling Alternative Fuels.

2.32.4. Street Sign Prices and Advertisements.

(a) The unit price must be in terms of price per kilogram in whole cents (e.g., $3.49 per kg, not $3.499 per kg).

(b) The sign or advertisement must include the service pressure (expressed in megapascals) at which the dispenser(s) delivers hydrogen fuel (e.g., H35 or H70).

(Added 2010)

2.33. Oil.

2.33.1. Labeling of Vehicle Engine (Motor) Oil. – Vehicle engine (motor) oil shall be labeled.

2.33.1.1. Viscosity. – The label on any vehicle engine (motor) oil container, receptacle, dispenser, or storage tank, and any invoice or receipt from service on an engine that includes the installation of vehicle engine (motor) oil dispensed from a receptacle, dispenser, or storage tank, shall contain the viscosity grade classification preceded by the letters "SAE" in accordance with SAE International's latest version of SAE J300, "Engine Oil Viscosity Classification."

2.33.1.2. Intended Use. – The label on any vehicle engine (motor) oil container shall contain a statement of its intended use in accordance with the latest version of SAE J183, "Engine Oil Performance and Engine Service Classification (other than Energy Conserving)."

2.33.1.3. Brand. – The label on any vehicle engine (motor) oil container and the invoice or receipt from service on an engine that includes the installation of vehicle engine (motor) oil dispensed from a receptacle, dispenser, or storage tank shall contain the name, brand, trademark, or trade name of the vehicle engine (motor) oil.

2.33.1.4. Engine Service Category. – The label on any vehicle engine (motor) oil container, receptacle, dispenser, or storage tank and the invoice or receipt from service on an engine that includes the installation of vehicle engine (motor) oil dispensed from a receptacle, dispenser, or storage tank shall contain the engine service category, or categories, displayed in letters not less than 3.18 mm (1/8 in) in height, as defined by the latest version of SAE J183, "Engine Oil Performance and Engine Service Classification (other than Energy Conserving)" or API Publication 1509, "Engine Oil Licensing and Certification System."

2.33.1.4.1. Inactive or Obsolete Service Categories. – The label on any vehicle engine (motor) oil container, receptacle, dispenser, or storage tank and the invoice or receipt from service on an engine that includes the installation of vehicle engine (motor) oil dispensed from a receptacle, dispenser, or storage tank shall bear a plainly visible cautionary statement in compliance with the latest version of SAE J183, Appendix A, whenever the vehicle engine (motor) oil in the container or in bulk does not meet an active API service category as defined by the latest version of SAE J183, "Engine Oil Performance and Engine Service Classification (other than Energy Conserving)."

2.33.1.5. Tank Trucks or Rail Cars. – Tank trucks, rail cars, and other types of delivery trucks that are used to deliver vehicle engine (motor) oil are not required to display the SAE viscosity grade and service category or categories on such tank trucks, rail cars, and other types of delivery trucks.

(Amended 2013)

2.33.1.6. Documentation. – When the engine (motor) oil is sold in bulk, an invoice, bill of lading, shipping paper, or other documentation must accompany each delivery. This document must identify the quantity of engine (motor) oil delivered as defined in Sections 2.33.1.1. Viscosity; 2.33.1.2. Intended Use; 2.33.1.3. Brand; 2.33.1.4. Engine Service Category; the name and address of the seller and buyer; and the date and time of the sale. For inactive or obsolete service categories, the documentation shall also bear a plainly visible cautionary statement as required in Section 2.33.1.4.1. Inactive or Obsolete Service Categories, documentation must be retained at the retail establishment for a period of not less than one year.

(Added 2013)

(Added 2012) (Amended 2013)

2.34. Retail Sales of Electricity Sold as a Vehicle Fuel.

2.34.1. Definitions.

2.34.1.1. Electricity Sold as Vehicle Fuel. – Electrical energy transferred to and/or stored onboard an electric vehicle primarily for the purpose of propulsion.

2.**34.1.2. Electric Vehicle Supply Equipment (EVSE).** – The conductors, including the ungrounded, grounded, and equipment grounding conductors; the electric vehicle connectors; attachment plugs; and all other fittings, devices, power outlets, or apparatuses installed specifically for the purpose of measuring, delivering, and computing the price of electrical energy delivered to the electric vehicle.

2.34.1.3. Fixed Service. – Service that continuously provides the nominal power that is possible with the equipment as it is installed.

2.34.1.4. Variable Service. – Service that may be controlled resulting in periods of reduced, and/or interrupted transfer of electrical energy.

2.34.1.5. Nominal Power. – Refers to the "intended" or "named" or "stated" as opposed to "actual" rate of transfer of electrical energy (i.e., power).

2.34.2. Method of Sale. – All electrical energy kept, offered, or exposed for sale and sold at retail as a vehicle fuel shall be in units in terms of the megajoule (MJ) or kilowatt-hour (kWh). In addition to the fee assessed for the quantity of electrical energy sold, fees may be assessed for other services; such fees may be based on time measurement and/or a fixed fee.

2.34.3. Retail Electric Vehicle Supply Equipment (EVSE) Labeling.

(a) A computing EVSE shall display the unit price in whole cents (e.g., $0.12) or tenths of one cent (e.g., $0.119) on the basis of price per megajoule (MJ) or kilowatt-hour (kWh). In cases where the electrical energy is unlimited or free of charge, this fact shall be clearly indicated in place of the unit price.

(b) For fixed service applications, the following information shall be conspicuously displayed or posted on the face of the device:

(1) the level of EV service expressed as the nominal power transfer (i.e., nominal rate of electrical energy transfer), and

(2) the type of electrical energy transfer (e.g., AC, DC, wireless).

(c) For variable service applications, the following information shall be conspicuously displayed or posted on the face of the device:

(1) the type of delivery (i.e., variable);

(2) the minimum and maximum power transfer that can occur during a transaction, including whether service can be reduced to zero;

(3) the condition under which variations in electrical energy transfer will occur; and

(4) the type of electrical energy transfer (e.g., AC, DC, wireless).

(d) Where fees will be assessed for other services in direct connection with the fueling of the vehicle, such as fees based on time measurement and/or a fixed fee, the additional fees shall be displayed.

(e) The EVSE shall be labeled in accordance with 16 CFR, Part 309 – FTC Labeling Requirements for Alternative Fuels and Alternative Fueled Vehicles.

(f) The EVSE shall be listed and labeled in accordance with the National Electric Code® (NEC) NFPA 70, Article 625 Electric Vehicle Charging Systems (**www nfpa.org**).

2.34.4. Street Sign Prices and Other Advertisements. – Where electrical energy unit price information is presented on street signs or in advertising other than on EVSE:

(a) The electrical energy unit price shall be in terms of price per megajoule (MJ) or kilowatt-hour (kWh) in whole cents (e.g., $0.12) or tenths of one cent (e.g., $0.119). In cases where the electrical energy is unlimited or free of charge, this fact shall be clearly indicated in place of the unit price.

(b) In cases where more than one electrical energy unit price may apply over the duration of a single transaction to sales to the general public, the terms and conditions that will determine each unit price and when each unit price will apply shall be clearly displayed.

(c) For fixed service applications, the following information shall be conspicuously displayed or posted:

(1) the level of EV service expressed as the nominal power transfer (i.e., nominal rate of electrical energy transfer), and

(2) the type of electrical energy transfer (e.g., AC, DC, wireless).

(d) For variable service applications, the following information shall be conspicuously displayed or posted:

(1) the type of delivery (i.e., variable);

(2) the minimum and maximum power transfer that can occur during a transaction, including whether service can be reduced to zero;

(3) the conditions under which variations in electrical energy transfer will occur; and

(4) the type of electrical energy transfer (e.g., AC, DC, wireless).

Where fees will be assessed for other services in direct connection with the fueling of the vehicle, such as fees based on time measurement and/or a fixed fee, the additional fees shall be included on all street signs or other advertising.

(Added 2013)

Section 3. General

3.1. Presentation of Price. – Whenever an advertised, posted, or labeled price per unit of weight, measure, or count for any commodity includes a fraction of a cent, all elements of the fraction shall be prominently displayed, and the numerals expressing the fraction shall be immediately adjacent to, of the same general design and style as, and at least one half the height and width of, the numerals representing the whole cent.

(Added 1976)

3.2. Allowable Differences: Combination Quantity Declarations. – Whenever the method of sale for a bulk or packaged commodity requires the use of a statement that includes count in addition to weight, measure, or size, the following shall apply to the particular commodity:

3.2.1. Beverageware: Pressed and Blown Tumblers and Stemware. – The allowable difference between actual and declared capacity shall be:

(a) **SI Units:**

(1) plus or minus 10 mL for items of 200 mL capacity or less; and

(2) plus or minus 5 % of the stated capacity for items over 200 mL capacity.

(Added 1973) (Amended 1974, 1979, and 1980)

(b) **Inch-pound Units:**

(1) plus or minus ¼ fl oz for items of 5 fl oz capacity or less; and

(2) plus or minus 5 % of the stated capacity for items over 5 fl oz capacity.

3.3. Labeling of Machines that Dispense Packaged Commodities. – All vending machines dispensing packaged commodities shall indicate:

(a) product identity;

(b) net quantity; and

(c) the party responsible for the vending machine.

Examples:
"For service or refunds contact: the XYZ Cola Company, Rockville, MD 20800; Telephone: (301) 555-1000," or "See attendant inside for refunds."

(Amended 1995)

(d) the requirements for product identity and net quantity can be met either by display of the package or by information posted on the outside of the machine.

(Added 1972)

3.4. Railroad Car Tare Weights. – Whenever stenciled tare weights on freight cars are employed in the sale of commodities or the assessment of freight charges, the following conditions and requirements shall apply:

3.4.1. Newly Stenciled Tare Weights. – All newly stenciled or re-stenciled tare weights shall be accurately represented to the nearest 50 kg for metric units and to the nearest 100 lb for inch-pound units, and the representation shall include the date of weighing.

(Amended 1979)

3.4.2. Allowable Difference. – The allowable difference between actual tare weight and stenciled tare weight on freight cars in use shall be per Section 3.4.2.(a) SI allowable difference or Section 3.4.2.(b) Inch-pound allowable difference.

(a) **SI allowable difference:**

(1) plus or minus 150 kg for cars 25 000 kg or less;

(2) plus or minus 200 kg for cars over 25 000 kg to and including 30 000 kg; and

(3) plus or minus 250 kg for cars over 30 000 kg.

(Added 1979)

(b) **Inch-pound allowable difference:**

(1) plus or minus 300 lb for cars 50 000 lb or less;

(2) plus or minus 400 lb for cars over 50 000 lb to and including 60 000 lb; and

(3) plus or minus 500 lb for cars over 60 000 lb.

3.4.3. Verification or Change of Tare Weights. – Tare weight determinations for verification or change of stenciled weights shall only be made on properly prepared and adequately cleaned freight cars.

3.4.4. Special Cars. – Tank cars, covered hopper cars, flat cars equipped with multi-deck racks or special superstructure, mechanical refrigerator cars, and house type cars equipped with special lading protective devices must be reweighed and re-stenciled only by owners or their authorized representatives:

(a) when car bears no lightweight (empty weight) stenciling; and

(b) when repairs or alterations result in a change of weight in excess of the permissible lightweight tolerance.

(Added 1974) (Renumbered 1985)

(Added 1973) (Amended 1974, 1979, and 1985)

Section 4. Revocation of Conflicting Regulations

All provisions of all orders and regulations heretofore issued on this same subject that are contrary to or inconsistent with the provisions of this regulation, and specifically __________, are hereby revoked.

(Added 1971)

Section 5. Effective Date

This regulation shall become effective on _____________.

Given under my hand and the seal of my office in the City of __________ on this __________ day of __________.

Signed ____

(Added 1971) (Amended 1973)

C. Uniform Unit Pricing Regulation

as adopted by
The National Conference on Weights and Measures*

1. Background

The Uniform Unit Pricing Regulation (UPR) (renamed in 1983) provides a national approach to the subject for those jurisdictions choosing to adopt such a regulation. The traditional approach of the Conference in drafting Uniform Regulations has been to design specific implementing Regulations for the enforcement of the broader requirements of the Uniform Weights and Measures Law. Given the authority of Sections 12.(c) and (d), and the mandate of Section 16. of this Law, as well as the trend in unit pricing, both voluntary and mandatory, the UPR is considered appropriate. Unit pricing has been a concern of the weights and measures official and has been required for random weight packages for a long time.

In 1993, the NCWM was contacted by several weights and measures jurisdictions and retail trade associations who requested that the UPR be updated to add new commodity groups and pricing requirements. The comments indicated that many commodity groups for non-food products were not included in the table and that some of the required units may not be appropriate for many of the new products being sold in stores. Another concern was that the UPR specified pricing only on the basis of price per pound on most products sold by weight. This has resulted in some jurisdictions not enforcing the requirements on stores that voluntarily unit price on the basis of price per ounce instead of price per pound. The NCWM agreed that the UPR should be revised to encourage wider adoption and use of the uniform regulation and that provisions for unit pricing in metric units should be included.

At the 1997 Annual Meeting, the NCWM adopted revisions to the regulation to permit retail stores that voluntarily provide unit pricing to present prices using various units of measure.

The NCWM eliminated the table of product groupings because it is difficult to keep it current and inclusive, so some newer products were not included under the uniform requirements. The table was replaced with requirements that specify that the unit price is to be based on price per ounce or pound, or price per 100 grams or kilogram, if the packaged commodity is labeled by weight. For example, the proposed revisions would require the unit price for soft drinks sold in various package sizes (e.g., 12 fl oz cans through 2 L bottles) to be uniformly and consistently displayed in terms of either price per fluid ounce, price per quart, or price per liter. The NCWM also increased the price of commodities exempted from unit pricing from 10 cents to 50 cents. The NCWM believed these revisions would ensure that unit pricing information facilitates value comparison between different package sizes and/or brands offered for sale in a store.

The NCWM also considered several comments on this item from members of the U.S. Metric Association (USMA). Most of these comments suggested that the UPR be amended to require unit pricing in metric units and permit inch-pound unit pricing to be provided voluntarily. When it developed the proposed revisions, the NCWM included guidelines for both inch-pound and metric unit pricing and believes this is the correct approach to implementing metric revisions in the regulation. The NCWM would like to make it clear that the UPR applies only when stores voluntarily provide unit pricing information. Its purpose is to provide a standard that retailers must follow to ensure that consumers will have pricing information that helps them make value comparisons. The decision to provide unit price information in metric or inch-pound units rests with retailers who will respond to consumer preference. The NCWM believes that consumer preference will be the deciding factor as to when and how quickly metric unit pricing is used in the marketplace. Therefore, the NCWM does not support amendments to include mandatory provisions in the UPR as these provisions would take the decision to go to metric unit pricing out of the hands of consumers and retailers. Finally, the NCWM does not want to include any requirement that may discourage retailers from voluntarily providing unit price information.

(Amended 1997)

**The National Conference on Weights and Measures (NCWM) is supported by the National Institute of Standards and Technology (NIST) in partial implementation of its statutory responsibility for "cooperation with the states in securing uniformity in weights and measures laws and methods of inspection."*

2. Status of Promulgation

The table beginning on page 10 shows the status of adoption of the Uniform Unit Pricing Regulation.

Uniform Unit Pricing Regulation

Table of Contents

Section **Page**

THIS PAGE INTENTIONALLY LEFT BLANK

Uniform Unit Pricing Regulation

Section 1. Application

Except for random and uniform weight packages that clearly state the unit price in accordance with existing regulations, any retail establishment providing unit price information for packaged commodities shall provide the unit price information in the manner prescribed herein.

Section 2. Terms for Unit Pricing

The declaration of the unit price of a particular commodity in all package sizes offered for sale in a retail establishment shall be uniformly and consistently expressed in terms of:

(a) Price per kilogram or 100 g, or price per pound or ounce, if the net quantity of contents of the commodity is in terms of weight.

(b) Price per liter or 100 mL, or price per dry quart or dry pint, if the net quantity of contents of the commodity is in terms of dry measure or volume.

(c) Price per liter or 100 mL, or price per gallon, quart, pint, or fluid ounce, if the net quantity of contents of the commodity is in terms of liquid volume.

(d) Price per individual unit or multiple units if the net quantity of contents of the commodity is in terms of count.

(e) Price per square meter, square decimeter, or square centimeter, or price per square yard, square foot, or square inch, if the net quantity of contents of the commodity is in terms of area.

Section 3. Exemptions

(a) Small Packages. – Commodities shall be exempt from these provisions when packaged in quantities of less than 28 g (1 oz) or 29 mL (1 fl oz) or when the total retail price is 50 cents or less.

(b) Single Items. – Commodities shall be exempt from these provisions when only one brand in only one size is offered for sale in a particular retail establishment.

(c) Infant Formula. – For "infant formula," unit price information may be based on the reconstituted volume. "Infant formula" means a food that is represented for special dietary use solely as a food for infants by reason of its simulation of human milk or suitability as a complete or partial substitute for human milk.

(d) Variety and Combination Packages. – Variety and Combination Packages as defined in Section 2.9 and Section 2.10 in the Uniform Packaging and Labeling Regulation [*NOTE 1*, page 143] shall be exempt from these provisions.

NOTE 1: *See "Uniform Packaging and Labeling Regulation."*

Section 4. Pricing

(a) The unit price shall be to the nearest cent when a dollar or more.

(b) If the unit price is under a dollar, it shall be listed:

(1) to the tenth of a cent; or

(2) to the whole cent.

The retail establishment shall have the option of using (b)(1) or (b)(2), but shall not implement both methods.

The retail establishment shall accurately and consistently use the same method of rounding up or down to compute the price to the whole cent.

Section 5. Presentation of Price

(a) In any retail establishment in which the unit price information is provided in accordance with the provisions of this regulation, that information may be displayed by means of a sign that offers the unit price for one or more brands and/or sizes of a given commodity, by means of a sticker, stamp, sign, label, or tag affixed to the shelf upon which the commodity is displayed, or by means of a sticker, stamp, sign, label, or tag affixed to the consumer commodity.

(b) Where a sign providing unit price information for one or more sizes or brands of a given commodity is used, that sign shall be displayed clearly and in a non-deceptive manner in a central location as close as practical to all items to which the sign refers.

(c) If a single sign or tag includes the unit price information for more than one brand or size of a given commodity, the following information shall be provided:

(1) the identity and the brand name of the commodity;

(2) the quantity of the packaged commodity, if more than one package size per brand is displayed;

(3) the total retail sales price; and

(4) the price per appropriate unit, in accordance with Section 2. Terms for Unit Pricing.

Section 6. Uniformity

(a) If different brands or package sizes of the same consumer commodity are expressed in more than one unit of measure (e.g., soft drinks are offered for sale in 2 L bottles and 12 fl oz cans), the retail establishment shall unit price the items consistently.

(b) When metric units appear on the consumer commodity in addition to other units of measure, the retail establishment may include both units of measure on any stamps, tags, labels, signs, or lists.

Section 7. Effective Date

This regulation shall become effective on __________, 20__.

Given under my hand and the seal of my office in the City of __________ on this __________ day of ________, 20__.

Signed ____
(Amended 1997)

D. Uniform Regulation for the Voluntary Registration of Servicepersons and Service Agencies for Commercial Weighing and Measuring Devices

as adopted by
The National Conference on Weights and Measures*

1. Background

The Uniform Regulation covering the registration of servicepersons and service agencies was developed and adopted by the National Conference on Weights and Measures (NCWM) in 1966, retitled in 1983, and substantially revised in 1984. It is designed to promote uniformity among those jurisdictions that provide for or are contemplating the establishment of some type of control over the servicing of commercial weighing and measuring devices. It offers to a serviceperson or to a service agency the opportunity to register and carries with it the privilege of restoring devices to service and of placing new or used devices in service.

Two unique features of the registration plan are its voluntary nature and the provision for reciprocity. Registration is not required; however, the privileges gained make it attractive. Also, in order to provide maximum effectiveness of the program and to reduce legal obstacles to a minimum to service across state lines, provision is made for reciprocity in certification of standards and testing equipment among states.

2. Status of Promulgation

The table beginning on page 10 shows the status of adoption of the Uniform Regulation for the Voluntary Registration of Servicepersons and Service Agencies for Commercial Weighing and Measuring Devices.

**The National Conference on Weights and Measures (NCWM) is supported by the National Institute of Standards and Technology (NIST) in partial implementation of its statutory responsibility for "cooperation with the states in securing uniformity in weights and measures laws and methods of inspection."*

THIS PAGE INTENTIONALLY LEFT BLANK

Uniform Regulation for the Voluntary Registration of Servicepersons and Service Agencies for Commercial Weighing and Measuring Devices

Table of Contents

Section **Page**

THIS PAGE INTENTIONALLY LEFT BLANK

Uniform Regulation for the Voluntary Registration of Servicepersons and Service Agencies for Commercial Weighing and Measuring Devices

Section 1. Policy

For the benefit of the users, manufacturers, and distributors of commercial weighing and measuring devices, it shall be the policy of the Director of Weights and Measures, hereinafter referred to as "Director," to accept registration of (a) an individual and (b) an agency providing acceptable evidence that he, she, or it is fully qualified by training or experience to install, service, repair, or recondition a commercial weighing or measuring device; has a thorough working knowledge of all appropriate weights and measures laws, orders, rules, and regulations; and has possession of, or has available for use, and will use suitable and calibrated weights and measures field standards and testing equipment appropriate in design and adequate in amount. (An employee of the government shall not be eligible for registration.)

The Director will check the qualifications of each applicant. It will be necessary for an applicant to have available sufficient field standards and equipment (see Section 5, Minimum Equipment).

It shall also be the policy of the Department to issue a "Certificate of Registration" to qualified applicants whose applications for registration are approved. This Certificate grants authority to remove rejection seals and tags placed on Commercial and Law Enforcement Weighing and Measuring Devices by authorized weights and measures officials, to place in service repaired devices that were rejected, and to place in service devices that have been newly installed.

The Director is NOT guaranteeing the work or fair dealing of a Registered Serviceperson or Service Agency. He will, however, remove from the registration list any Registered Serviceperson or Service Agency that performs unsatisfactory work or takes unfair advantage of a device owner.

Registration with the Director shall be on a voluntary basis. The Director shall reserve the right to limit or reject the application of any Serviceperson or Service Agency and to revoke his, her, or its permit to remove rejection seals or tags for good cause.

This policy shall in no way preclude or limit the right and privilege of any individual or agency not registered with the Director to install, service, repair, or recondition a commercial weighing or measuring device (see Section 7, Privileges and Responsibilities of a Voluntary Registrant).

(Added 1966) (Amended 1984 and 2005)

Section 2. Definitions

2.1. Registered Serviceperson. – Any individual who for hire, award, commission, or any other payment of any kind, installs, services, repairs, or reconditions a commercial weighing or measuring device, and who voluntarily registers with the Director of Weights and Measures.

(Added 1966)

2.2. Registered Service Agency. – Any agency, firm, company, or corporation that for hire, award, commission, or any other payment of any kind installs, services, repairs, or reconditions a commercial weighing or measuring device, and that voluntarily registers with the Director of Weights and Measures. Under agency registration, identification of individual servicepersons shall be required.

(Added 1966) (Amended 1984)

2.3. Commercial and Law Enforcement Weighing and Measuring Devices. – Any weight or measure or weighing or measuring device commercially used or employed in establishing the size, quantity, extent, area, or measurement of quantities, things, produce, or articles for distribution or consumption, purchased, offered, or submitted for sale, hire, or award, or in computing any basic charge or payment for services rendered on the basis of weight or measure. It shall also include any accessory attached to or used in connection with a commercial weighing or measuring device when such accessory is so designed or installed that its operation affects the accuracy of the device. It also includes weighing and measuring equipment in official use for the enforcement of law or for the collection of statistical information by government agencies.

(Added 1966) (Amended 1984)

Section 3. Registration Fee

There shall be charged by the Director an annual fee of $__________ per Registered Serviceperson and $__________ per Registered Service Agency to cover costs at the time application for registration is made, and annually, thereafter.

(Added 1966) (Amended 1984)

Section 4. Voluntary Registration

An individual or agency qualified by training or experience may apply for registration to service weighing devices or measuring devices on an application form supplied by the Director. Said form, duly signed and witnessed, shall include certification by the applicant that the individual or agency is fully qualified to install, service, repair, or recondition whatever devices for the service of which competence is being registered; has in possession or available for use, and will use, all necessary testing equipment and standards; and has full knowledge of all appropriate weights and measures laws, orders, rules, and regulations. An applicant also shall submit appropriate evidence or references as to qualifications. Application for registration shall be voluntary, but the Director is authorized to reject or limit any application.

(Added 1966) (Amended 1984)

Section 5. Minimum Equipment

Applicants must have available sufficient standards and equipment to adequately test devices as set forth in the Notes section of each applicable code in NIST Handbook 44, "Specifications, Tolerances, and Other Technical Requirements for Weighing and Measuring Devices." This equipment will meet the specifications of NIST 105-series standards (or other suitable and designated standards). This section shall not preclude the use of additional field standards and/or equipment, as approved by the Director, for uniform evaluation of device performance (see Section 9, Examination and Calibration or Certification of Standards and Testing Equipment).

(Added 1984) (Amended 2005)

Section 6. Certificate of Registration

The Director will review and check the qualifications of each applicant. The Director shall issue to the applicant a "Certificate of Registration," including an assigned registration number if it is determined that the applicant is qualified. The "Certificate of Registration" will expire 1 year from the date of issuance.

(Added 1966) (Amended 1984)

Section 7. Privileges and Responsibilities of a Voluntary Registrant

A bearer of a Certificate of Registration shall have the authority to remove an official rejection tag or mark placed on a weighing or measuring device by the authority of the Director; place in service, until such time as an official examination can be made, a weighing or measuring device that has been officially rejected; and place in service, until such time as an official examination can be made, a new or used weighing or measuring device. The registered serviceperson or service agency is responsible for installing, repairing, and adjusting devices such that the devices are adjusted as closely as practicable to zero error.

(Added 1966) (Amended 1984)

Section 8. Placed in Service Report

The Director shall furnish each registered serviceperson and registered service agency with a supply of report forms to be known as "Placed in Service Reports." Such a form shall be executed in triplicate, shall include the assigned registration number, and shall be signed by a registered serviceperson or by a serviceperson representing a registered agency for each rejected device restored to service and for each newly installed device placed in service. Within 24 hours after a device is restored to service or placed in service, the original of the properly executed Placed in Service Report, together with any official rejection tag removed from the device, shall be forwarded to the Director at ________(address). The duplicate copy of the report shall be handed to the owner or operator of the device, and the triplicate copy of the report shall be retained by the Registered Serviceperson or Registered Service Agency.

(Added 1966) (Amended 2005)

Section 9. Examination and Calibration or Certification of Standards and Testing Equipment

All field standards that are used for servicing and testing weights and measures devices for which competence is registered shall be submitted to the Director for initial and subsequent verification and calibration at intervals determined by the Director. A registered serviceperson or registered service agency shall not use in servicing commercial weighing or measuring devices any field standards or testing equipment that have not been calibrated or verified by the Director. In lieu of submission of physical standards, the Director may accept calibration and/or verification reports from any laboratory that is formally accredited or recognized. The Director shall maintain a list of organizations from which the state will accept calibration reports. The state shall retain the right to periodically monitor calibration results and/or to verify field standard compliance to specifications and tolerances when field standards are initially placed into service or at any intermediate point between calibrations.

(Added 1966) (Amended 1984, 1999, and 2005)

Section 10. Revocation of Certificate of Registration

The Director is authorized to suspend or revoke a Certificate of Registration for good cause which shall include, but not be limited to: taking of unfair advantage of an owner of a device; failure to have test equipment or standards certified; failure to use adequate testing equipment; or failure to adjust commercial or law enforcement devices to comply with Handbook 44 subsequent to service or repair.

(Added 1966) (Amended 1984)

Section 11. Publication of Lists of Registered Servicepersons and Registered Service Agencies

The Director shall publish, from time to time as he deems appropriate, and may supply upon request, lists of Registered Servicepersons and Registered Service Agencies.

(Added 1966)

Section 12. Effective Date

This regulation shall become effective on __________.

(Added 1966)

THIS PAGE INTENTIONALLY LEFT BLANK

E. Uniform Open Dating Regulation

as adopted by
The National Conference on Weights and Measures*

1. Background

Numerous state and local jurisdictions have provided for, or are considering, mandatory open dating of certain packaged commodities. Additionally, many commodities in the marketplace are now voluntarily open dated. Lack of uniformity between jurisdictions could impede the orderly flow of commerce.

In 1985, the National Conference on Weights and Measures (NCWM), in concert with the Association of Food and Drug Officials, wrote a new Uniform Regulation. It resolved the differences in the versions developed independently by the two organizations.

The regulation provides two options for implementation by the states. One requires open dating on all perishable foods and the other permits voluntary open dating of such foods. In the latter (voluntary) case, the open dating must then conform to the uniform regulation. Notes to Sections 1.1. Purpose and 3.1. "Sell By" Date indicate the alternate wording for the voluntary version of the Regulation.

2. Status of Promulgation

The table beginning on page 10 shows the status of adoption of the Uniform Open Dating Regulation.

**The National Conference on Weights and Measures (NCWM) is supported by the National Institute of Standards and Technology (NIST) in partial implementation of its statutory responsibility for "cooperation with the states in securing uniformity in weights and measures laws and methods of inspection."*

THIS PAGE INTENTIONALLY LEFT BLANK

Uniform Open Dating Regulation

Table of Contents

Section **Page**

THIS PAGE INTENTIONALLY LEFT BLANK

Uniform Open Dating Regulation [NOTE 1, page 157]

Section 1. Purpose, Scope, and Application

1.1. Purpose. [NOTE 1, page 157] – The purpose of this regulation is to prescribe mandatory uniform date labeling of prepackaged, perishable foods and to prescribe optional uniform date labeling that must be used whenever a packager elects to use date labeling on prepackaged foods that are not perishable. Open dating is intended for use and understanding by both distributors and consumers when judging food qualities.

NOTE 1: *Alternatively, this regulation may be adopted to require uniformity of open dating of perishable foods whenever a packager voluntarily elects to use date labeling. In such instances Sections 1.1. Purpose and 3.1. "Sell By" Date are reworded in the following manner:*

> ***1.1. Purpose.*** *– The purpose of this regulation is to prescribe uniform date labeling that must be used whenever a packager elects to use date labeling on a prepackaged food. Open date labeling is intended for use and understanding by both distributors and consumers when judging food qualities.*
>
> ***3.1. "Sell By" Date.*** *– If a retail food establishment elects to sell or offer for sale a prepackaged perishable food identified with a "sell by" date, the "sell by" date used must be as prescribed by this regulation.*

1.2. Scope and Application. – This regulation prescribes the manner of date labeling, the method of determining the appropriate date, required records, responsible persons, and the foods subject to this regulation. This regulation provides for the permissible sale of a regulated food after the expiration of the date on the label. This regulation does not apply to any food that is not prepackaged or is exempted by Section 8.

Section 2. Definitions

2.1. "Sell By" Date. – A recommended last date of sale that permits a subsequent period before deterioration of qualities described in 2.2. Perishable Food, 2.3. Semi Perishable Food, and 2.4. Long Shelf Life Food.

2.2. Perishable Food. – Any food having a significant risk of spoilage, loss of value, or loss of palatability within 60 days of the date of packaging.

2.3. Semi Perishable Food. – Any food for which a significant risk of spoilage, loss of value, or loss of palatability occurs only after a minimum of 60 days, but within 6 months, after the date of packaging.

2.4. Long Shelf Life Food. – Any food for which a significant risk of spoilage, loss of value, or loss of palatability does not occur sooner than 6 months after the date of packaging, including foods preserved by freezing, dehydrating, or being placed in a hermetically sealed container.

2.5. Prepackaged. – Food packaged prior to being displayed or offered for retail sale.

2.6. "Best If Used By" Date. – A date prior to deterioration of qualities described in 2.3. Semi Perishable Food and 2.4. Long Shelf Life Food.

2.7. Person. – An individual, partnership, association, or corporation.

Section 3. Sale of Perishable Food and Date Determination

3.1. "Sell By" Date. [NOTE 1, page 157] – A retail food establishment shall not sell or offer for sale a prepackaged perishable food unless it is identified with a "sell by" date as prescribed by this regulation.

3.2. Sale After Expiration of "Sell By" Date.

3.2.1. Advertisement. – Perishable food shall not be offered for sale after the "sell by" date unless it is wholesome and advertised in a conspicuous manner as being offered for sale after the recommended last date of sale. The placement of a sign, sticker, or tag is acceptable for such advertising if it is easily readable and clearly identifies the perishable food as having passed the recommended last date of sale.

3.2.2. Responsibility for Advertisement. – The retailer or final seller is responsible for the advertisement, described in Section 3.2.1. Advertisement, of a perishable food offered for sale after the recommended last date of sale.

3.3. Determination of "Sell By" Date.

3.3.1. Reasonable Period for Consumption. – A manufacturer, processor, packer, re-packer, retailer, or other person who prepackages perishable food, shall determine a date that allows a reasonable period after sale for consumption of the food without physical spoilage, loss of value, or loss of palatability. A reasonable period for consumption shall consist of at least one third of the approximate total shelf life of the perishable food.

3.3.2. Responsibility for "Sell By" Date. – A retailer who purchases prepackaged perishable food may upon written agreement with the person prepackaging such food determine, identify, and be responsible for the "sell by" date placed on or attached to each package of such food.

3.4. Manner of Expressing Date.

3.4.1. Month and Day, or Day of Week. – A person described in Section 3.3.1. Reasonable Period for Consumption or 3.3.2. Responsibility for "Sell By" Date shall place or attach to each package of perishable food a date by month and day. However, bakery products with a shelf life of not more than seven days may be dated with the day of the week representing the last recommended day of sale.

3.4.2. The term "Sell By". – The "sell by" date shall be displayed with the term "sell by" or words of similar import immediately preceding or immediately over the designated date unless a prominent notice is on the label describing the date as a "sell by" date and indicating the location of the date.

3.4.3. Abbreviation of Weekday. – If the day of the week is solely designated as provided in Section 3.4.1. Month and Day, or Day of Week the name of the day may be abbreviated by the use of either the first two or first three letters of the name of the day.

3.4.4. Expression of Month and Day. – Except as provided for in Section 3.4.1. Month and Day, or Day of Week the date shall be designated by:

(a) the first three letters of the month, preceded or followed by a numeral indicating the calendar day; or

(b) the month represented numerically followed by a numeral designation of the calendar day.

The month and day designation shall be separated by a period, slash, dash, or spacing. When a numeral designation of the first nine days of the month is used, the number shall include a zero as the first digit; for example, 01 or 03.

(Amended 1987)

3.4.5. Expression of the Year. – The "sell by" date may include the year following the day if such year is expressed as a two or four digit number separated as described in Section 3.4.4. Expression of Month and Day.

Section 4. Sale of Semi Perishable and Long Shelf Life Food

4.1. "Best If Used By" Date. – A manufacturer, processor, packer, re-packer, or other person who prepackages semi perishable or long shelf life food may place upon or attach to the package an open date providing it is designated by the "best if used by" date.

4.2. Sale After Expiration of "Best If Used By" Date. – A retail food establishment may sell or offer for sale food beyond the designated "best if used by" date provided the food is wholesome and the sensory physical quality standards for that food have not significantly diminished.

4.3. Manner of Expressing Date. – The "best if used by" date as required by Section 4.1. "Best If Used By" Date shall be placed upon or attached to each container or package and be limited to the terms "best if used by" or words of similar import followed by or immediately over the date designated by the month and year unless a prominent notice is on the label describing the date as a "best if used by" date and indicating the location of the date. The date shall be designated by the first three letters of the month followed by a numeral indicating the year. The use of the day of the month is permissible provided that the day of the month is placed prior to the month; for example, 30 Jun 81.

Section 5. Placement of the Date

The date, whether "sell by" or "best if used by," shall be printed, stamped, embossed, perforated, or otherwise shown on the package, label on the package, or tag attached to the package in a manner that is easily readable and separate from other information, graphics, or lettering so as to be clearly visible to a prospective purchaser. The date shall not be superimposed on other required information or obscured by other information, graphics, or pricing. Regardless of the type size used, the date shall be easily readable. These requirements do not preclude a supplemental notice elsewhere on a package describing and/or indicating the location of the date.

Section 6. Factors for the Date Determination

A person who, as provided for in this regulation, places either the "sell by" date or "best if used by" date on a package shall determine the date by taking into consideration the food quality, characteristics, formulation, processing impact, packaging or container and other protective wrapping or coating, customary transportation, and storage and display conditions. For purposes of calculating this date, home storage conditions shall be considered to be similar to those in the usual retail store except that the date for refrigerated food may be calculated by using a home storage temperature standard of 40 °F (4.4 °C).

Section 7. Records

A person who is responsible for establishing the date for perishable, semi perishable, and long shelf life food shall keep a record of the method used to determine the date. A record revision is necessary whenever a factor affecting date determination is altered. Such record shall be retained for not less than six months after the most recent "sell by" or "best if used by" date and shall be available during normal business hours for examination upon request by __________ (agency name).

Section 8. Exemptions

8.1. This regulation does not apply to perishable fruits or vegetables in a container permitting sensory examination.

8.2. This regulation does not apply to prepackaged perishable foods open dated according to requirements of federal law or regulation.

Section 9. Preemption of Local, County, and Municipal Ordinance

A municipality or county shall not adopt or impose standards or requirements other than those provided for in this regulation.

Section 10. Effective Date

This regulation shall become effective on and after __________.

F. Uniform Regulation for National Type Evaluation

as adopted by
The National Conference on Weights and Measures*

1. Background

The Uniform Regulation for National Type Evaluation was adopted by the NCWM at the 68th Annual Meeting in 1983 and is a necessary adjunct to recognize and enable participation in the National Type Evaluation Program administered by the National Conference on Weights and Measures (NCWM). The Regulation specifically authorizes: type evaluation; recognition of a NCWM "Certificate of Conformance" of type; the State Measurement Laboratory to operate as a Participating Laboratory, if authorized by the National Institute of Standards and Technology (NIST) under its program of recognition of State Measurement Laboratories; and, the state to charge fees to those persons who seek type evaluation of weighing and measuring devices.

(Amended 2000)

At the 81st Annual Meeting in 1996, the NCWM adopted major revisions to the Uniform Regulation for National Type Evaluation. These revisions were made to clarify the requirements and incorporate the policies and guidelines adopted by the Executive Committee as published in NCWM Publication 14, "Technical Policy, Checklists, and Test Procedures."

(Amended 1997)

2. Intent

It is the intent of this regulation to have all states use the National Type Evaluation, as approved by the NCWM, as their examining procedure. If a state does not wish to establish a Participating Laboratory, Section 2.4. Participating Laboratory and Section 4. Participating Laboratory may be deleted.

3. Status of Promulgation

The table beginning on page 10 shows the status of adoption of the Uniform Regulation for National Type Evaluation.

**The National Conference on Weights and Measures (NCWM) is supported by the National Institute of Standards and Technology in partial implementation of its statutory responsibility for "cooperation with the states in securing uniformity in weights and measures laws and methods of inspection."*

THIS PAGE INTENTIONALLY LEFT BLANK

Uniform Regulation for National Type Evaluation

Table of Contents

THIS PAGE INTENTIONALLY LEFT BLANK

Uniform Regulation for National Type Evaluation

Section 1. Application

This regulation shall apply to [*NOTE 1*, page 165] any type of device and/or equipment covered in National Institute of Standards and Technology (NIST) Handbook 44 for which evaluation procedures have been published in the National Conference on Weights and Measures (NCWM), Publication 14, "National Type Evaluation Program, Technical Policy, Checklists, and Test Procedures."

NOTE 1: *This section can be amended to include a list of devices or device types to which NTEP evaluation criteria does not apply. Additionally, a state can amend this section to allow it to conduct a type evaluation and issue a "Certificate of Approval." This approach should be limited to occasions where formal NTEP Type Evaluation criteria does not apply and to new technologies or device applications where the development of criteria is deemed necessary by the Director.*

Section 2. Definitions

2.1. Active Certificate of Conformance (CC). – A document issued based on testing by a Participating Laboratory, which the certificate holder maintains in active status under the National Type Evaluation Program (NTEP). The document constitutes evidence of conformance of a type with the requirements of this document, NIST Handbook 44, "Specifications, Tolerances, and Other Technical Requirements for Weighing and Measuring Devices," and the test procedures contained in NCWM Publication 14. By maintaining the Certificate in active status, the Certificate holder declares the intent to continue to manufacture or remanufacture the device consistent with the type and in conformance with the applicable requirements. A device is traceable to an active CC if: (a) it is of the same type identified on the Certificate, and (b) it was manufactured during the period that the Certificate was maintained in active status. For manufacturers of grain moisture meters, maintenance of active status also involves annual participation in the NTEP Laboratory On-going Calibration Program, OCP (Phase II).

(Amended 2000, 2001, and 2004)

2.2. Device. – A piece of commercial or law enforcement equipment as defined in Section 2.15. Commercial and Law Enforcement Equipment. A device may be a single unit or a combination of separate and compatible main elements. A device shall include, at a minimum, those main elements that: (a) perform the measurement, and (b) process the measurement signals up to the first indicated or recorded value of the final quantity upon which the transaction is based.

(Amended 2004)

2.3. Director. – Means the __________ of the Department of __________.

2.4. Manufactured Device. – Any commercial weighing or measuring device shipped as new from the original equipment manufacturer.

(Added 2001)

2.5. National Type Evaluation Program. – A program of cooperation between the NCWM, NIST, other federal agencies, the states, and the private sector for determining, on a uniform basis, conformance of a type with the relevant provisions of National Institute of Standards and Technology Handbook 44, "Specifications, Tolerances, and Other Technical Requirements for Weighing and Measuring Devices," and NCWM, Publication 14, "National Type Evaluation Program, Technical Policy, Checklists, and Test Procedures.".

(Amended 2000)

2.6. One-of-a-Kind Device. – A (non-NTEP) device designed to meet unique demands for a specific installation and of a specific design which is not commercially available elsewhere (one such device per manufacturer). If a device manufactured for sale by a company has been categorized and tested as a "one-of-a-kind" device and the manufacturer then decides to manufacture an additional device or devices of that same type, the device will no longer be considered a "one-of-a-kind." This also applies to a device that has been determined to be a "one-of-a-kind" device by a weights and measures jurisdiction in one state and the manufacturer decides to manufacture and install another device of that same type in another state. In this case, the manufacturer of the device must request an NTEP evaluation on the device through the normal application process unless NTEP has already deemed that such evaluation will not be conducted.

(Amended 1998)

2.7. Participating Laboratory. – Any State Measurement Laboratory or State Weights and Measures Agency or other laboratory that has been authorized to conduct a type evaluation under the NTEP.

(Amended 2001)

2.8. Person. – The term "person" means both singular and plural, as the case demands, and includes individuals, partnerships, corporations, companies, societies, and associations.

2.9. Remanufactured Device. – A device that is disassembled, checked for wear, parts replaced or fixed, reassembled, and made to operate like a new device of the same type.

(Amended 2001)

2.10. Remanufactured Element. – An element that is disassembled, checked for wear, parts replaced or fixed, reassembled, and made to operate like a new element of the same type.

(Added 2001)

2.11. Repaired Device. – A device on which work is performed that brings the device back into proper operating condition.

(Amended 2001)

2.12. Repaired Element. – An element on which work is performed that brings the element back into proper operating condition.

(Added 2001)

2.13. Type. – A model or models of a particular device, measurement system, instrument, or element that positively identifies the design. A specific type may vary in its measurement ranges, size, performance, and operating characteristics as specified in the CC.

2.14. Type Evaluation. – The testing, examination, and/or evaluation of a type by a Participating Laboratory under the NTEP.

2.15. Commercial and Law Enforcement Equipment.

(a) Weighing and measuring equipment commercially used or employed in establishing the size, quantity, extent, area, or measurement of quantities, things, produce, or articles for distribution or consumption, purchased, offered, or submitted for sale, hire, or award, or in computing any basic charge or payment for services rendered on the basis of weight or measure.

(b) Any accessory attached to or used in connection with a commercial weighing or measuring device when such accessory is so designed that its operation affects the accuracy of the device.

(c) Weighing and measuring equipment in official use for the enforcement of law or for the collection of statistical information by government agencies. [*NOTE 2*, page 166]

NOTE 2: *The section is identical to G-A.1. Commercial and Law Enforcement Equipment, Section 1.10. General Code, National Institute of Standards and Technology Handbook 44 for definition of "commercial" and "law enforcement equipment."*

Section 3. Certificate of Conformance (CC)

The Director shall require a device to be traceable to an active Certificate of Conformance (CC) prior to its installation or use for commercial or law enforcement purposes. If the device consists of separate and compatible main elements, each main element shall be traceable to a CC. A device is traceable to a CC if:

(a) it is of the same type identified on the Certificate; and

(b) it was manufactured during the period that the Certificate was maintained in active status.

(Amended 2001 and 2004)

Section 4. Prohibited Acts and Exemptions

(a) Except for a device exempted by this section, no person shall sell a device unless it is traceable to an active CC.

(Amended 2001)

(b) Except for a device exempted by this section, no person shall use a device unless it is traceable to an active CC.

(Amended 2001)

(c) A device in service in this State prior to __________, ____, (date) that meets the specifications, tolerances, and other technical requirements of National Institute of Standards and Technology Handbook 44 shall not be required to be traceable to an active CC.

(Amended 2001)

(d) A device in service in this State prior to __________, ____, (date) removed from service by the owner or on which the department has issued a removal order after __________, ____, (date) and returned to service at a later date shall be modified to meet all specifications, tolerances, and other technical requirements of National Institute of Standards and Technology Handbook 44 effective on the date of the return to service. Such a device shall not be required to be traceable to an active CC.

(Amended 2001)

(e) A device in service in this State prior to __________, ____, (date) which is repaired after such date shall meet the specifications, tolerances, and other technical requirements of National Institute of Standards and Technology Handbook 44 and shall not be required to be traceable to an active CC.

(Amended 2001)

(f) A device in service in this State prior to __________, ____, (date) that is still in use may be installed at another location in this state provided the device meets requirements in effect as of the date of installation in the new location; however, the device shall not be required to be traceable to an active CC.

(Amended 2001)

(g) A device in service in another State prior to __________, ____, (date) may be installed in this State; however, the device shall meet the specifications, tolerances, and technical requirements for weighing and measuring devices in National Institute of Standards and Technology Handbook 44 and be traceable to an active CC.

(Amended 2001)

(h) **One-of-a-kind Device.** – The Director may accept the design of a one-of-a-kind device without an NTEP evaluation pending inspection and performance testing to satisfy that the device complies with Handbook 44 and is capable of performing within the Handbook 44 requirements for a reasonable period of time under normal conditions of use. Indicators and load cells in all "one-of-a-kind" scale installations must have an active NTEP CC as evidence that the system meets the influence factor requirements of Handbook 44.

(Amended 1998 and 2001)

(i) **Repaired Device.** – If a person makes changes to a device to the extent that the metrological characteristics are changed, that specific device is no longer traceable to the active CC.

(Amended 2001)

(j) **Remanufactured Device.** – If a person repairs or remanufactures a device, they are obligated to repair or remanufacture it consistent with the manufacturer's original design; otherwise, that specific device is no longer traceable to an active CC.

(Amended 2001)

(k) **Copy of a Device.** – The manufacturer who copies the design of a device that is traceable to an active CC, but which is made by another company, must obtain a separate CC for the device. The CC for the original device shall not apply to the copy.

(l) **Device Components.** – If a person buys a load cell(s) and an indicating element that are traceable to CCs and then manufactures a device from the parts, that person shall obtain an active CC for the device.

(Amended 2001)

Section 5. Participating Laboratory and Agreements

The Director is authorized to:

(a) Operate a Participating Laboratory as part of the NTEP. In this regard, the Director is authorized to charge and collect fees for type evaluation services.

(b) Cooperate with and enter into agreements with any person in order to carry out the purposes of the act.

Section 6. Revocation of Conflicting Regulations

All provisions of all orders and regulations before issued on this same subject that are contrary to or inconsistent with the provisions of this regulation, are hereby revoked.
(Amended 2001)

Section 7. Effective Date

This regulation shall become effective on __________.
(Amended 2001)

G. Uniform Engine Fuels and Automotive Lubricants Regulation

as adopted by
The National Conference on Weights and Measures*

1. Background

In 1984, the National Conference on Weights and Measures (NCWM) adopted a Section 2.20. in the Uniform Regulation for the Method of Sale of Commodities requiring that motor fuels containing alcohol be labeled to disclose to the retail purchaser that the fuel contains alcohol. The delegates deemed this action necessary since motor vehicle manufacturers were qualifying their warranties with respect to some gasoline-alcohol blends, motor fuel users were complaining to weights and measures officials about fuel quality and vehicle performance, and ASTM International (ASTM) had not yet finalized quality standards for oxygenated (which includes alcohol-containing) fuels. While a few officials argued weights and measures officials should not cross the line from quantity assurance programs to programs regulating quality, the delegates were persuaded that the issue needed immediate attention.

A Motor Fuels Task Force was appointed in 1984 to develop mechanisms for achieving uniformity in the evaluation and regulation of motor fuels. The Task Force developed the Uniform Motor Fuel Inspection Law (see the Uniform Engine Fuels and Automotive Lubricants Inspection Law section of this handbook) and the Uniform Engine Fuel and Automotive Lubricants Regulation to accompany the law. The Uniform Law required registration and certification of motor fuel as meeting ASTM standards. The regulation defined the ASTM standards to be applied to motor fuel.

In 1992, the NCWM established the Petroleum Subcommittee under the Laws and Regulations Committee. The subcommittee recommended major revisions to the Regulation that was adopted at the 80th NCWM in 1995. The scope of the regulation was expanded to include all engine fuels, petroleum products, and automotive lubricants; its title was changed accordingly; and the fuel specifications and method of sale sections were revised to address the additional products. Other changes included expansion of the definitions section and addition of sections on retail storage tanks, condemned product, registration of engine fuels designed for special use, and test methods and reproducibility limits.

In 2007, the Petroleum Subcommittee (now referred to as the Fuels and Lubricants Subcommittee) undertook a review of this regulation to update it by eliminating reference to "petroleum products" and to reflect the addition of new engine fuels to the marketplace.

At the 2008 NCWM Interim Meeting, the Laws and Regulations Committee changed the Petroleum Subcommittee's name to the Fuels and Lubricants Subcommittee (FALS) in recognition of its work with a wide variety of fuels including petroleum and biofuels.

2. Status of Promulgation

The Uniform Regulation for Engine Fuels and Automotive Lubricants was adopted by the NCWM in 1995 and the latest amendments were adopted in 2008. The status of state actions with respect to this Regulation is shown in the table beginning on page 10.
(Amended 2008)

**The National Conference on Weights and Measures (NCWM) is supported by the National Institute of Standards and Technology (NIST) in partial implementation of its statutory responsibility for "cooperation with the states in securing uniformity in weights and measures laws and methods of inspection."*

THIS PAGE INTENTIONALLY LEFT BLANK

Uniform Engine Fuels and Automotive Lubricants Regulation

Table of Contents

Uniform Engine Fuels and Automotive Lubricants Regulation

Section 1. Definitions

1.1. ASTM International. (www.astm.org) – The international voluntary consensus standards organization formed for the development of standards on characteristics and performance of materials, products, systems, and services, and the promotion of related knowledge.

1.2. Antiknock Index (AKI). – The arithmetic average of the Research Octane Number (RON) and Motor Octane Number (MON): AKI = (RON+MON)/2. This value is called by a variety of names, in addition to antiknock index, including: octane rating, posted octane, (R+M)/2 octane.

1.3. Automatic Transmission Fluid. – A product intended for use in a passenger vehicle, other than a bus, as either lubricant, coolant, or liquid medium in any type of fluid automatic transmission that contains a torque converter. For the purposes of this regulation, fluids intended for use in continuously variable transmissions are not considered "Automatic Transmission Fluid."

(Added 2004)

1.4. Automotive Fuel Rating. – The automotive fuel rating required under the amended Octane Certification and Posting Rule (or as amended, the Fuel Rating Rule), 16 CFR Part 306. Under this Rule, sellers of liquid automotive fuels, including alternative fuels, must determine, certify, and post an appropriate automotive fuel rating. The automotive fuel rating for gasoline is the antiknock index (octane rating). The automotive fuel rating for alternative liquid fuels consists of the common name of the fuel, along with a disclosure of the amount, expressed as a minimum volume percent of the principal component of the fuel. For alternative liquid automotive fuels, a disclosure of other components, expressed as a minimum volume percent, may be included, if desired.

1.5. Automotive Gasoline, Automotive Gasoline-Oxygenate Blend. – A type of fuel suitable for use in spark ignition automobile engines and also commonly used in marine and non-automotive applications.

1.6. Aviation Gasoline. – A type of gasoline suitable for use as a fuel in an aviation spark-ignition internal combustion engine.

1.7. Aviation Turbine Fuel. – A refined middle distillate suitable for use as a fuel in an aviation gas turbine internal combustion engine.

1.8. Base Gasoline. – All components other than ethanol in a blend of gasoline and ethanol.

1.9. Biodiesel. – A fuel comprised of mono-alkyl esters of long chain fatty acids derived from vegetable oils or animal fats, designated B100.

1.10. Biodiesel Blend. – A fuel comprised of a blend of biodiesel fuel with petroleum-based diesel fuel, designated BXX. In the abbreviation BXX, (e.g., B20) represents the volume percentage of biodiesel fuel in the blend.

1.11. Cetane Number. – A numerical measure of the ignition performance of a diesel fuel obtained by comparing it to reference fuels in a standardized engine test.

1.12. Compressed Natural Gas (CNG). – Natural gas which has been compressed and dispensed into fuel storage containers and is suitable for use as an engine fuel.

1.13. Denatured Fuel Ethanol. – "Ethanol" as defined in Section 1.20. Ethanol.

1.14. Diesel Fuel. – A refined middle distillate suitable for use as a fuel in a compression-ignition (diesel) internal combustion engine.

1.15. Distillate. - Any product obtained by condensing the vapors given off by boiling petroleum or its products.

1.16. EPA. – The United States Environmental Protection Agency (**www.epa.gov**).

1.17. E85 Fuel Ethanol. – A blend of ethanol and hydrocarbons of which the ethanol portion is nominally 75 to 85 volume percent denatured fuel ethanol.

1.18. Engine Fuel. – Any liquid or gaseous matter used for the generation of power in an internal combustion engine.

1.19. Engine Fuels Designed for Special Use. – Engine fuels designated by the Director as requiring registration. These fuels normally do not have ASTM or other national consensus standards applying to their quality or usability; common special fuels are racing fuels and those intended for agricultural and other off-road applications.

1.20. Ethanol. – Also known as "Denatured Fuel Ethanol," means nominally anhydrous ethyl alcohol meeting ASTM D4806 standards. It is intended to be blended with gasoline for use as a fuel in a spark-ignition internal combustion engine. The denatured fuel ethanol is first made unfit for drinking by the addition of the Alcohol and Tobacco Tax and Trade Bureau (TTB), **www.ttb.gov**, approved substances before blending with gasoline.

1.21. Fuel Cell. – An electrochemical energy conversion device in which fuel and an oxidant react to generate electricity without consumption, physically or chemically, of its electrodes or electrolytes.

(Added 2012)

1.22. Fuel Oil. – Refined oil middle distillates, heavy distillates, or residues of refining, or blends of these, suitable for use as a fuel for heating or power generation, the classification of which shall be defined by ASTM D396.

1.23. Gasoline. – A volatile mixture of liquid hydrocarbons generally containing small amounts of additives suitable for use as a fuel in a spark-ignition internal combustion engine.

1.24. Gasoline-Alcohol Blend. – A fuel consisting primarily of gasoline and a substantial amount (more than 0.35 mass percent of oxygen, or more than 0.15 mass percent of oxygen if methanol is the only oxygenate) of one or more alcohols.

1.25. Gasoline Gallon Equivalent (GGE). – Equivalent to 2.567 kg (5.660 lb) of natural gas.

1.26. Gasoline Liter Equivalent (GLE). – Equivalent to 0.678 kg (1.495 lb) of natural gas.

1.27. Gasoline-Oxygenate Blend. – A fuel consisting primarily of gasoline along with a substantial amount (more than 0.35 mass percent of oxygen, or more than 0.15 mass percent of oxygen if methanol is the only oxygenate) of one or more oxygenates.

1.28. Gear Oil. – An oil used to lubricate gears, axles, or some manual transmissions.

(Added 2004)

1.29. Hydrogen Fuel. – A fuel composed of molecular hydrogen intended for consumption in a surface vehicle or electricity production device with an internal combustion engine or fuel cell.

(Added 2012)

1.30. Internal Combustion Engine. – A device used to generate power by converting chemical energy bound in the fuel via spark-ignition or compression ignition combustion into mechanical work to power a vehicle or other device.

(Added 2012)

1.31. Kerosene. – (or "Kerosine") A refined middle distillate suitable for use as a fuel for heating or illuminating, the classification of which shall be defined by the latest version of ASTM D3699, "Standard Specification for Kerosine."

1.32. Lead Substitute. – An EPA-registered gasoline additive suitable, when added in small amounts to fuel, to reduce or prevent exhaust valve recession (or seat wear) in automotive spark-ignition internal combustion engines designed to operate on leaded fuel.

1.33. Lead Substitute Engine Fuel. – For labeling purposes, a gasoline or gasoline-oxygenate blend that contains a "lead substitute".

1.34. Leaded. – For labeling purposes, any gasoline or gasoline-oxygenate blend which contains more than 0.013 g of lead per liter (0.05 g lead per U.S. gal). NOTE: EPA defines leaded fuel as one which contains more than 0.0013 g of phosphorus per liter (0.005 g per U.S. gal), or any fuel to which lead or phosphorus is intentionally added.

1.35. Liquefied Natural Gas (LNG). – Natural gas that has been liquefied at – 126.1 °C (– 259 °F) and stored in insulated cryogenic tanks for use as an engine fuel.

1.36. Liquefied Petroleum Gas (LPG). – A mixture of normally gaseous hydrocarbons, predominantly propane, or butane, or both, that has been liquefied by compression or cooling, or both to facilitate storage, transport, and handling.

1.37. Low Temperature Operability. – A condition which allows the uninterrupted operation of a diesel engine through the continuous flow of fuel throughout its fuel delivery system at low temperatures. Fuels with adequate low temperature operability characteristics have the ability to avoid wax precipitation and clogging in fuel filters.

(Added 1998) (Amended 1999)

1.38. Lubricant. – Oil (See 1.44. below.).

(Added 2008)

1.39. Lubricity. – A qualitative term describing the ability of a fluid to affect friction between, and wear to, surfaces **al** in relative motion under load.

(Added 2003)

1.40. M85 Fuel Methanol. – A blend of methanol and hydrocarbons of which the methanol portion is nominally 70 to 85 volume percent.

1.41. Motor Octane Number. – A numerical indication of a spark-ignition engine fuel's resistance to knock obtained by comparison with reference fuels in a standardized ASTM D2700, "Motor Method Engine Test."

1.42. Motor Oil. – An oil that reduces friction and wear between the moving parts within a reciprocating internal combustion engine and also serves as a coolant. For the purposes of this regulation, "vehicle motor oil" refers to motor oil which is intended for use in light- to heavy-duty vehicles including cars, sport utility vehicles, vans, trucks, buses, and off-road farming and construction equipment. For the purposes of this regulation, "recreational motor oil" refers to motor oil which is intended for use in four-stroke cycle engines used in motorcycles, ATVs, and lawn and garden equipment. For the purposes of this regulation, motor oil also means engine oil.

(Added 2004)

1.43. MTBE. – Methyl tertiary-butyl ether.

(Added 2008)

1.44. Oil. – A motor oil, engine oil, and/or gear oil.

(Added 2004)

1.45. Oxygen Content of Gasoline. – The percentage of oxygen by mass contained in a gasoline.

1.46. Oxygenate. – An oxygen-containing, ashless, organic compound, such as an alcohol or ether, which can be used as a fuel or fuel supplement.

1.47. Reformulated Gasoline (RFG). – A gasoline or gasoline-oxygenate blend certified to meet the specifications and emission reduction requirements established by the Clean Air Act Amendments of 1990, as amended by the Energy Policy Act of 2005, required to be sold for use in automotive vehicles in extreme and severe ozone non-attainment areas and those areas which opt to require reformulated gasoline.

(Amended 2008)

1.48. Research Octane Number. – A numerical indication of a spark-ignition engine fuel's resistance to knock obtained by comparison with reference fuels in a standardized ASTM D2699, "Research Method Engine Test."

1.49. SAE (SAE International). – A technical organization for engineers, scientists, technicians, and others who cooperate closely in the engineering, design, manufacture, use, and maintainability of self-propelled vehicles.

1.50. Substantially Similar. – Refers to the EPA's "Substantially Similar" rule, Section 211 (f) (1) of the Clean Air Act [42 U.S.C. 7545 (f) (1)].

1.51. ThermStability. – The ability of a fuel to resist the thermal stress which is experienced by the fuel when exposed to high temperatures in a fuel delivery system. Such stress can lead to formation of insoluble gums or organic particulates. Insolubles (e.g., gums or organic particulates) can clog fuel filters and contribute to injector deposits.

(Added 1998) (Amended 1999)

1.52. Unleaded. – When used in conjunction with "engine fuel" or "gasoline" means any gasoline or gasoline-oxygenate blend to which no lead or phosphorus compounds have been intentionally added and which contains not more than 0.013 g of lead per liter (0.05 g lead per U.S. gallon) and not more than 0.0013 g of phosphorus per liter (0.005 g phosphorus per U.S. gallon).

1.53. Wholesale Purchaser Consumer. – Any person who is an ultimate gasoline consumer of fuel methanol, fuel ethanol, diesel fuel, biodiesel, fuel oil, kerosene, aviation turbine fuels, natural gas, compressed natural gas, or liquefied petroleum gas and who purchases or obtains the product from a supplier and receives delivery of that product into a storage tank.

(Added 1998) (Amended 1999)

Section 2. Standard Fuel Specifications

2.1. Gasoline and Gasoline-Oxygenate Blends.

2.1.1. Gasoline and Gasoline-Oxygenate Blends (as defined in this regulation). – Shall meet the latest version of ASTM D4814, "Standard Specification for Automotive Spark-Ignition Engine Fuel" except for the permissible offsets for ethanol blends as provided in Section 2.1.2. Gasoline-Ethanol Blends.

(Added 2009)

2.1.2. Gasoline-Ethanol Blends. – When gasoline is blended with ethanol, the ethanol shall meet the latest version of ASTM D4806, "Standard Specification for Denatured Fuel Ethanol for Blending with Gasolines for Use as Automotive Spark-Ignition Engine Fuel," and the blend shall meet the latest version of ASTM D4814, "Standard Specification for Automotive Spark-Ignition Engine Fuel," with the following permissible exceptions:

(a) The maximum vapor pressure shall not exceed the ASTM D4814 limits by more than:

(1) 1.0 psi for blends containing 9 to 10 volume percent ethanol from June 1 through September 15.

(2) 1.0 psi for blends containing one or more volume percent ethanol for volatility classes A, B, C, D from September 16 through May 31.

(3) 0.5 psi for blends containing one or more volume percent ethanol for volatility Class E from September 16 through May 31.

The vapor pressure exceptions in subsections 2.1.2. Gasoline-Ethanol Blends will remain in effect until May 1, 2016, or until ASTM incorporates changes to the vapor pressure maximums for ethanol blends, whichever occurs earlier.

NOTE 1: *The temperature values (e.g., 54 °C, 50. °C, 41.5 °C) are presented in the format prescribed in ASTM E29 "Standard Practice for Using Significant Digits in Test Data to Determine Conformance with Specifications."*

(Added 2009) (Amended 2012)

2.1.3. Minimum Antiknock Index (AKI). – The AKI shall not be less than the AKI posted on the product dispenser or as certified on the invoice, bill of lading, shipping paper, or other documentation;

2.1.4. Minimum Motor Octane Number. – The minimum motor octane number shall not be less than 82 for gasoline with an AKI of 87 or greater;

2.1.5. Minimum Lead Content to Be Termed "Leaded." – Gasoline and gasoline oxygenate blends sold as "leaded" shall contain a minimum of 0.013 g of lead per liter (0.05 g per U.S. gallon);

2.1.6. Lead Substitute Gasoline. – Gasoline and gasoline-oxygenate blends sold as "lead substitute" gasoline shall contain a lead substitute which provides protection against exhaust valve seat recession equivalent to at least 0.026 g of lead per liter (0.10 g per U.S. gallon).

2.1.6.1. Documentation of Exhaust Valve Seat Protection. – Upon the request of the Director, the lead substitute additive manufacturer shall provide documentation to the Director that demonstrates that the treatment level recommended by the additive manufacturer provides protection against exhaust valve seat recession equivalent to or better than 0.026 g/L (0.1 g/gal) lead. The Director may review the documentation and approve the lead substitute additive before such additive is blended into gasoline. This documentation shall consist of:

(a) test results as published in the Federal Register by the EPA Administrator as required in Section 211(f)(2) of the Clean Air Act; or

(b) until such time as the EPA Administrator develops and publishes a test procedure to determine the additive's effectiveness in reducing valve seat wear, test results and description of the test procedures used in comparing the effectiveness of 0.026 g per liter lead and the recommended treatment level of the lead substitute additive shall be provided.

2.1.7. Blending. – Leaded, lead substitute, and unleaded gasoline-oxygenate blends shall be blended according to the EPA "substantially similar" rule or an EPA waiver for unleaded fuel.

(Amended 2009)

2.2. Diesel Fuel. – Shall meet the latest version of ASTM D975, "Standard Specification for Diesel Fuel Oils."

2.2.1. Premium Diesel Fuel. – All diesel fuels identified on retail dispensers, bills of lading, invoices, shipping papers, or other documentation with terms such as premium, super, supreme, plus, or premier must conform to the following requirements:

(a) **Cetane Number.** – A minimum cetane number of 47.0 as determined by the latest version ASTM D613, "Standard Test Method for Cetance Nuber of Diesel Fuel Oil".

(b) **Low Temperature Operability.** – A cold flow performance measurement which meets the latest version of ASTM D975, "Standard Specification for Diesel Fuel Oils," tenth percentile minimum ambient air temperature charts and maps by either ASTM Standard Test Method D2500 (Cloud Point) or the latest version of ASTM Standard D4539, "Low Temperature Flow Test, LTFT." Low temperature operability is only applicable October 1 - March 31 of each year.

(c) **Thermal Stability.** – A minimum reflectance measurement of 80 % as determined by the latest version ASTM Standard Test Method D6468 (180 min, 150 °C).

(d) **Lubricity.** – A maximum wear scar diameter of 520 microns as determined by the latest version ASTM D6079, "Standard Test Method for Evaluating Lubricity of Diesel Fuels by the High-Frequency Reciprocating Rig (HFRR)." If an enforcement jurisdiction's single test of more than 560 microns is determined, a second test shall be conducted. If the average of the two tests is more than 560 microns, the sample does not conform to the requirements of this part.

(Amended 2003)

2.3. Aviation Turbine Fuels. – Shall meet the latest version of ASTM D1655, "Standard Specification for Aviation Turbine Fuels."

2.4. Aviation Gasoline. – Shall meet the most recent version of one of the following, as appropriate:

(a) **ASTM D910** – "Standard Specification for Aviation Gasoline;" or

(b) **ASTM D6227** – "Standard Specification for Grade 82 Unleaded Aviation Gasoline."

(Amended 2008)

2.5. Fuel Oils. – Shall meet the latest version of ASTM D396, "Standard Specification for Fuel Oils."

2.6. Kerosene (Kerosine). – Shall meet the latest version of ASTM D3699, "Standard Specification for Kerosine."

2.7. Ethanol. – Intended for blending with gasoline shall meet the latest version of ASTM D4806, "Standard Specification for Denatured Fuel Ethanol for Blending with Gasolines for Use as Automotive Spark-Ignition Engine Fuel."

2.8. Liquefied Petroleum (LP) Gases. – Shall meet the latest version ASTM D1835, "Standard Specification for Liquefied Petroleum (LP) Gases."

NOTE: *Also reference Gas Processors Association 2140, Liquefied Petroleum Gas Specification and Test Methods.*

2.9. Compressed Natural Gas (CNG). – Shall meet the latest version of SAE J1616, "Recommended Practice for Compressed Natural Gas Vehicle Fuel."

2.10. E85 Fuel Ethanol. – Shall meet the latest version of ASTM D5798, "Standard Specification for Fuel Ethanol (Ed75-Ed85) for Automotive Spark-Ignition Engines."

(Added 1997)

2.11. M85 Fuel Methanol. – Shall meet the latest version of ASTM D5797, "Standard Specification for Fuel Methanol M70-M85 for Automotive Spark Ignition Engines."

(Added 1997)

2.12. Motor Oil. – Shall not be sold or distributed for use unless the product conforms to the following specifications:

(a) performance claims listed on the label shall be evaluated against the latest version of SAE J183, "Engine Oil Performance and Engine Service Classification (Other than "Energy Conserving," API 1509 "Engine Oil Licensing and Certifications System," or other industry standards as applicable;

(b) the product shall meet its labeled viscosity grade specification as specified in the latest version of SAE J300, "Engine Oil Viscosity Classification; and

(c) any engine oil that is represented as "energy conserving" shall meet the requirements established by the latest version of SAE J1423, "Classification of Energy Conserving Engine Oil for Passenger Cars, Vans, Sport Utility Vehicles, and Light-Duty Trucks.

(Added 2004)

2.13. Products for Use in Lubricating Manual Transmissions, Gears, or Axles. – Shall not be sold or distributed for use in lubricating manual transmissions, gears, or axles unless the product conforms to the following specifications:

(a) it is labeled with one or more of the service designations found in the latest version of the SAE Information Report on axle and manual transmission lubricants, SAE J308, and API Publication 1560, and meets all applicable requirements of those designations;

(b) the product shall meet its labeled viscosity grade classification as specified in the latest version of SAE J306; and

(c) the product shall be free from water and suspended matter when tested by means of centrifuge, in accordance with the latest version of ASTM D2273, "Standard Test Method for Trace Sediment in Lubricating Oils."

(Added 2004)

2.14. Products for Use in Lubricating Automatic Transmissions. – Any automatic transmission fluid sold without limitation as to type of transmission for which it is intended shall meet all automotive manufacturers' recommended requirements for transmissions in general use in the state. Automatic transmission fluids that are intended for use only in certain transmissions, as disclosed on the label of its container, shall meet the latest automotive manufacturers' recommended requirements for those transmissions. Adherence to automotive manufacturers' recommended requirements shall be based on tests currently available to the lubricants' industry and the state regulatory agency. Any material offered for sale or sold as an additive to automatic transmission fluids shall be compatible with the automatic transmission fluid to which it is added, and shall meet all performance claims as stated on the label. Any manufacturer of any such product sold in this state shall provide, upon request by a duly authorized representative of the Director, documentation of any claims made on their product label.

(Added 2004)

2.15. Biodiesel. – B100 biodiesel intended for blending with diesel fuel shall meet the latest version of ASTM D6751, "Standard Specification for Biodiesel Fuel (B100) Blend Stock for Distillate Fuels."

(Added 2004)

2.16. Biodiesel Blends. – Blends of biodiesel and diesel fuels shall meet the following requirements:

(a) blends that contain less than or equal to 5 % must meet the latest version of ASTM D975, "Standard Specification for Diesel Fuel Oils";

(b) blends greater than 5 % biodiesel and that contain less than or equal to 20 % by volume shall meet the latest version of ASTM D7467, "Standard Specification for Diesel Fuel Oil, Biodiesel Blend (B6 to B20)";

(c) use of S15 biodiesel is required when blending into S15 low sulfur motor vehicle diesel fuel when the intention is to certify the fuel as S15 grade; and

(d) when blends greater than 20 % are offered for sale, the diesel fuel used in the blend shall meet the latest version of ASTM D975, "Standard Specification for Diesel Fuel Oils," and the biodiesel blend stock shall meet the specifications of paragraph 2.15. Biodiesel.

(Added 2004) (Amended 2008)

2.17. Hydrogen Fuel. – Shall meet the latest version of SAE J2719, "Hydrogen Fuel Quality for Fuel Cell Vehicles."

(Added 2012)

Section 3. Classification and Method of Sale of Petroleum Products

3.1. General Considerations.

3.1.1. Documentation. – When products regulated by this rule are sold, an invoice, bill of lading, shipping paper, or other documentation must accompany each delivery other than a retail sale. This document must identify the quantity, the name of the product, the particular grade of the product, the applicable automotive fuel rating, and oxygenate type and content (if applicable), the name and address of the seller and buyer, and the date and time of the sale. Documentation must be retained at the retail establishment for a period not less than one year.

(Amended 2008)

3.1.2. Retail Dispenser Labeling. – All retail dispensing devices must identify conspicuously the type of product, the particular grade of the product, and the applicable automotive fuel rating.

3.1.3. Grade Name. – The sale of any product under any grade name that indicates to the purchaser that it is of a certain automotive fuel rating or ASTM grade shall not be permitted unless the automotive fuel rating or grade indicated in the grade name is consistent with the value and meets the requirements of Section 2, Standard Fuel Specifications.

3.2. Automotive Gasoline and Automotive Gasoline-Oxygenate Blends.

3.2.1. Posting of Antiknock Index Required. – All automotive gasoline and automotive gasoline-oxygenate blends shall post the antiknock index in accordance with applicable regulations, 16 CFR Part 306 issued pursuant to the Petroleum Marketing Practices Act, as amended.

3.2.2. When the Term "Leaded" May be Used. – The term "leaded" shall be used only when the fuel meets specification requirements of paragraph 2.1.5. Minimum Lead Content to be Termed "Leaded."

3.2.3. Use of Lead Substitute Must be Disclosed. – Each dispensing device from which gasoline or gasoline-oxygenate blends containing a lead substitute is dispensed shall display the following legend: "Contains Lead Substitute." The lettering of this legend shall not be less than 12.7 mm (½ in) in height and the color of the lettering shall be in definite contrast to the background color to which it is applied.

3.2.4. Nozzle Requirements for Leaded Fuel. – Each dispensing device from which gasoline or gasoline-oxygenate blends that contain lead in amounts sufficient to be considered "leaded" gasoline, or lead substitute engine fuel, is sold shall be equipped with a nozzle spout having a terminal end with an outside diameter of not less than 23.63 mm (0.930 in).

3.2.5. Prohibition of Terms. – It is prohibited to use specific terms to describe a grade of gasoline or gasoline-oxygenate blend unless it meets the minimum antiknock index requirement shown in Table 1. Minimum Antiknock Index Requirements.

Table 1.
Minimum Antiknock Index Requirements

Term	Minimum Antiknock Index	
	ASTM D4814 Altitude Reduction Areas IV and V	All Other ASTM D4814 Areas
Premium, Super, Supreme, High Test	90	91
Midgrade, Plus	87	89
Regular Leaded	86	88
Regular, Unleaded (alone)	85	87
Economy	--	86

(Table 1. Amended 1997)

3.2.6. Method of Retail Sale. – Type of Oxygenate must be disclosed. All automotive gasoline or automotive gasoline-oxygenate blends kept, offered, or exposed for sale, or sold at retail containing at least 1.5 mass percent oxygen shall be identified as "with" or "containing" (or similar wording) the predominant oxygenate in the engine fuel. For example, the label may read "contains ethanol" or "with methyl *tertiary*-butyl ether (MTBE)." The oxygenate contributing the largest mass percent oxygen to the blend shall be considered the predominant oxygenate. Where mixtures of only ethers are present, the retailer may post the predominant oxygenate followed by the phrase "or other ethers" or alternatively post the phrase "contains MTBE or other ethers." In addition, gasoline-methanol blend fuels containing more than 0.15 mass percent oxygen from methanol shall be identified as "with" or "containing" methanol. This information shall be posted on the upper 50 % of the dispenser front panel in a position clear and conspicuous from the driver's position in a type at least 12.7 mm (½ in) in height, 1.5 mm ($^1/_{16}$ in) stroke (width of type).

(Amended 1996)

3.2.7. Documentation for Dispenser Labeling Purposes. – The retailer shall be provided, at the time of delivery of the fuel, on an invoice, bill of lading, shipping paper, or other documentation, a declaration of the predominant oxygenate or combination of oxygenates present in concentrations sufficient to yield an oxygen content of at least 1.5 mass percent in the fuel. Where mixtures of only ethers are present, the fuel supplier may identify either the predominant oxygenate in the fuel (i.e., the oxygenate contributing the largest mass percent oxygen) or alternatively, use the phrase "contains MTBE or other ethers." In addition, any gasoline containing more than 0.15 mass percent oxygen from methanol shall be identified as "with" or "containing" methanol. This documentation is only for dispenser labeling purposes; it is the responsibility of any potential blender to determine the total oxygen content of the engine fuel before blending.

(Amended 1996)

3.2.8. EPA Labeling Requirements also Apply. – Retailers and wholesale purchaser-consumers of gasoline shall comply with the EPA pump labeling requirements for gasoline containing greater than 10 volume percent (v%) up to 15 volume percent (v%) ethanol (E15) under 40 CFR § 80.1501.

(Added 2012)

3.3. Diesel Fuel.

3.3.1. Labeling of Grade Required. – Diesel Fuel shall be identified by grades No. 1-D, No. 2-D, or No. 4-D.

3.3.2. EPA Labeling Requirements Also Apply. – Retailers and wholesale purchaser-consumers of diesel fuel shall comply with EPA pump labeling requirements for sulfur under 40 CFR § 80.570.

3.3.3. Delivery Documentation for Premium Diesel. – Before or at the time of delivery of premium diesel fuel, the retailer or the wholesale purchaser-consumer shall be provided on an invoice, bill of lading, shipping paper, or other documentation a declaration of all performance properties that qualifies the fuel as premium diesel fuel as required in Section 2.2.1. Premium Diesel Fuel.

(Added 1998) (Amended 1999)

3.3.4. Nozzle Requirements for Diesel Fuel. – Each dispensing device from which diesel fuel is sold at retail shall be equipped with a nozzle spout with a diameter that conforms to the latest version of SAE J285, "Dispenser Nozzle Spouts for Liquid Fuels Intended for Use with Spark Ignition and Compression Ignition Engines." (Enforceable effective July 1, 2013)

(Added 2012)

(Amended 1998, 1999, 2008, and 2012)

3.4. Aviation Turbine Fuels.

3.4.1. Labeling of Grade Required. – Aviation turbine fuels shall be identified by Jet A, Jet A 1, or Jet B.

3.4.2. NFPA Labeling Requirements also Apply. – Each dispenser or airport fuel truck dispensing aviation turbine fuels shall be labeled in accordance with the most recent edition of National Fire Protection Association (NFPA 407), Standard for Aircraft Fuel Servicing.

NOTE: *For example, NFPA 407, 2007 edition: Section 4.3.18 Product Identification Signs. Each aircraft fuel servicing vehicle shall have a sign on each side and the rear to indicate the product. The sign shall have letters at least 75 mm (3 in) high of color sharply contrasting with its background for visibility. It shall show the word "FLAMMABLE" and the name of the product carried, such as "JET A," "JET B," "GASOLINE," or "AVGAS." (**NOTE:** Refer to the most recent edition NFTA 407.)*

3.5. Aviation Gasoline.

3.5.1. Labeling of Grade Required. – Aviation gasoline shall be identified by Grade 80, Grade 91, Grade 100, or Grade 100LL, or Grade 82UL

(Amended 2008)

3.5.2. NFPA Labeling Requirements also Apply. – Each dispenser or airport fuel truck dispensing aviation gasoline shall be labeled in accordance with the most recent edition of National Fire Protection Association (NFPA) 407, Standard for Aircraft Fuel Servicing.

NOTE: *For example, NFPA 407, 2007 edition: Section 4.3.18 Product Identification Signs. Each aircraft fuel servicing vehicle shall have a sign on each side and the rear to indicate the product. The sign shall have letters at least 3 in (75 mm) high of color sharply contrasting with its background for visibility. It shall show the word "FLAMMABLE" and the name of the product carried, such as "JET A," "JET B," "GASOLINE," or "AVGAS." (**NOTE**: Refer to the most recent edition NFTA 407.)*

3.6. Fuel Oils.

3.6.1. Labeling of Grade Required. – Fuel Oil shall be identified by the grades of No. 1 S500, No. 1 S5000, No. 2 S500, No. 2 S5000, No. 4 (Light), No. 4, No. 5 (Light), No. 5 (Heavy), or No. 6.
(Amended 2008)

3.7. Kerosene (Kerosine).

3.7.1. Labeling of Grade Required. – Kerosene shall be identified by the grades No. 1-K or No. 2-K.

3.7.2. Additional Labeling Requirements. – Each retail dispenser of kerosene shall be labeled as 1-K Kerosene or 2-K. In addition, No. 2-K dispensers shall display the following legend:

"Warning - Not Suitable For Use In Unvented Heaters Requiring No. 1-K."

The lettering of this legend shall not be less than 12.7 mm (½ in) in height by 1.5 mm (1/16 in) stroke; block style letters and the color of lettering shall be in definite contrast to the background color to which it is applied.

3.8. E85 Fuel Ethanol.

3.8.1. How to Identify E85 Fuel Ethanol. – Fuel ethanol shall be identified as E85.

3.8.2. Labeling Requirements.

(a) Fuel ethanol shall be labeled with its automotive fuel rating in accordance with 16 CFR Part 306.

(b) A label shall be posted which states "For Use in Flexible Fuel Vehicles (FFV) Only." This information shall be clearly and conspicuously posted on the upper 50 % of the dispenser front panel in a type at least 12.7 mm (½ in) in height, 1.5 mm (1/16 in) stroke (width of type). A label shall be posted which states, "Consult Vehicle Manufacturer Fuel Recommendations," and shall not be less than 6 mm (¼ in) in height by 0.8 mm (1/32 in) stroke; block style letters and the color shall be in definite contrast to the background color to which it is applied.

(Amended 2007 and 2008)

3.9. M85 Fuel Methanol.

3.9.1. How to Identify M85 Fuel Methanol. – Fuel methanol shall be identified as M85.

Example: M85

3.9.2. Retail Dispenser Labeling.

(a) Fuel methanol shall be labeled with its automotive fuel rating in accordance with 16 CFR Part 306.

Example: M85 Methanol

(b) A label shall be posted which states "For Use in Vehicles Capable of Using M85 Only." This information shall be clearly and conspicuously posted on the upper 50 % of the dispenser front panel in a type of at least 12.7 mm (½ in) in height, 1.5 mm (1/16 in) stroke (width of type).

(Amended 2008)

3.10. Liquefied Petroleum Gas (LPG).

3.10.1. How LPG is to be Identified. – Liquefied petroleum gases shall be identified by grades Commercial Propane, Commercial Butane, Commercial PB Mixtures or Special-Duty Propane (HD5).

3.10.2. Retail Dispenser Labeling. – Each retail dispenser of LPGs shall be labeled as "Commercial Propane," "Commercial Butane," "Commercial PB Mixtures," or "Special-Duty Propane (HD5)."

3.10.3. Additional Labeling Requirements. – LPG shall be labeled with its automotive fuel rating in accordance with 16 CFR Part 306.

3.10.4. NFPA Labeling Requirements Also Apply. (Refer to the most recent edition of NFPA 58.)

3.11. Compressed Natural Gas (CNG).

3.11.1. How Compressed Natural Gas is to be Identified. – For the purposes of this regulation, compressed natural gas shall be identified by the term "Compressed Natural Gas" or "CNG."

3.11.2. Retail Sales of Compressed Natural Gas Sold as a Vehicle Fuel.

3.11.2.1. Method of Retail Sale. – All CNG kept, offered, or exposed for sale or sold at retail as a vehicle fuel shall be in terms of the gasoline liter equivalent (GLE) or gasoline gallon equivalent (GGE).

3.11.2.2. Retail Dispenser Labeling.

3.11.2.2.1. Identification of Product. – Each retail dispenser of CNG shall be labeled as "Compressed Natural Gas."

3.11.2.2.2. Conversion Factor. – All retail CNG dispensers shall be labeled with the conversion factor in terms of kilograms or pounds. The label shall be permanently and conspicuously displayed on the face of the dispenser and shall have either the statement "1 Gasoline Liter Equivalent (GLE) is equal to 0.678 kg of Natural Gas" or "1 Gasoline Gallon Equivalent (GGE) is equal to 5.660 lb of Natural Gas" consistent with the method of sale used.

3.11.2.2.3. Pressure. – CNG is dispensed into vehicle fuel containers with working pressures of 16 574 kPa, 20 684 kPa, or 24 821 kPa. The dispenser shall be labeled 16 574 kPa, 20 684 kPa, or 24 821 kPa corresponding to the pressure of the CNG dispensed by each fueling hose.

3.11.2.2.4. NFPA Labeling. – NFPA Labeling requirements also apply. (Refer to NFPA 52.)

3.11.3. Nozzle Requirements for CNG. – CNG fueling nozzles shall comply with ANSI/AGA/CGA NGV 1.

3.12. Liquefied Natural Gas (LNG).

3.12.1. How Liquefied Natural Gas is to be Identified. – For the purposes of this regulation, liquefied natural gas shall be identified by the term "Liquefied Natural Gas" or "LNG."

3.12.2. Labeling of Retail Dispensers of Liquefied Natural Gas Sold as a Vehicle Fuel.

3.12.2.1. Identification of Product. – Each retail dispenser of LNG shall be labeled as "Liquefied Natural Gas."

3.12.2.2. Automotive Fuel Rating. – LNG automotive fuel shall be labeled with its automotive fuel rating in accordance with 16 CFR Part 306.

3.12.2.3. NFPA Labeling. – NFPA Labeling requirements also apply. (Refer to NFPA 57.)

3.13. Oil.

3.13.1. Labeling of Vehicle Engine (Motor) Oil Required.

3.13.1.1. Viscosity. – The label on any vehicle engine (motor) oil container, receptacle, dispenser, or storage tank and the invoice or receipt from service on an engine that includes the installation of vehicle engine (motor) oil dispensed from a receptacle, dispenser, or storage tank shall contain the viscosity grade classification preceded by the letters "SAE" in accordance with the SAE International's latest version of SAE J300, "Engine Oil Viscosity Classification."

(Amended 2012)

3.13.1.2. Intended Use. – The label on any vehicle engine (motor) oil container shall contain a statement of its intended use in accordance with the latest version of SAE J183, "Engine Oil Performance and Engine Service Classification (Other than 'Energy Conserving')."

(Amended 2012)

3.13.1.3. Brand. – The label on any vehicle engine (motor) oil container and the invoice or receipt from service on an engine that includes the installation of vehicle engine (motor) oil dispensed from a receptacle, dispenser, or storage tank shall contain the name, brand, trademark, or trade name of the vehicle engine (motor) oil.

(Added 2012)

3.13.1.4. Engine Service Category. – The label on any vehicle engine (motor) oil container, receptacle, dispenser or storage tank and the invoice or receipt from service on an engine that includes the installation of vehicle engine (motor) oil dispensed from a receptacle, dispenser, or storage tank shall contain the engine service category, or categories, displayed in letters not less than 3.18 mm (1/8 in) in height, as defined by the latest version of SAE J183, "Engine Oil Performance and Engine Service Classification (other than 'Energy Conserving')" or API Publication 1509, Engine Oil Licensing and Certification System.

(Amended 2012)

3.13.1.4.1. Inactive or Obsolete Service Categories. – The label on any vehicle engine (motor) oil container, receptacle, dispenser, or storage tank and the invoice or receipt from service on an engine that includes the installation of vehicle engine (motor) oil dispensed from a receptacle, dispenser, or storage tank shall bear a plainly visible cautionary statement in compliance with the latest version of SAE J183, "Engine Oil Performance and Engine Service Classification (Other than 'Energy Conserving')" Appendix A, whenever the vehicle engine (motor) oil in the container or in bulk does not meet an active API service category as defined by the latest version of SAE J183, "Engine Oil Performance and Engine Service Classification (Other than 'Energy Conserving')."

(Added 2012)

3.13.1.5. Tank Trucks or Rail Cars. – Tank trucks, rail cars, and types of delivery trucks that are used to deliver vehicle engine (motor) oil are not required to display the SAE viscosity grade and service category or categories on such tank trucks, rail cars, and other types of delivery trucks.

(Added 2012) (Amend 2013)

3.13.1.6. Documentation. – When the engine (motor) oil is sold in bulk, an invoice, bill of lading, shipping paper, or other documentation must accompany each delivery. This document must identify the quantity of engine (motor) oil delivered as defined in Sections 3.13.1.1. Viscosity; 3.13.1.2. Intended Use; 3.13.1.3. Brand; 3.13.1.4. Engine Service Category; the name and address of the seller and buyer; and the date and time of the sale. For inactive or obsolete service categories, the documentation shall also bear a plainly visible cautionary statement as required in Section 3.13.1.4.1. Inactive or Obsolete Service Categories, documentation must be retained at the retail establishment for a period of not less than one year.

(Added 2013)

(Amended 2012 and 2013)

3.13.2. Labeling of Recreational Motor Oil.

3.13.2.1. Viscosity. – The label on each container of recreational motor oil shall contain the viscosity grade classification preceded by the letters "SAE" in accordance with the SAE International's latest version of SAE J300, "Engine Oil Viscosity Classification."

3.13.2.2. Intended Use. – The label on each container of recreational motor oil shall contain a statement of its intended use in accordance with the latest version of SAE J300, "Engine Oil Viscosity Classification."

3.13.3. Labeling of Gear Oil.

3.13.3.1. Viscosity. – The label on each container of gear oil shall contain the viscosity grade classification preceded by the letters "SAE" in accordance with the SAE International's latest version of SAE J306, "Automotive Gear Lubricant Viscosity Classification" or SAE J300, "Engine Oil Viscosity Classification."

3.13.3.1.1. Exception. – Some automotive equipment manufacturers may not specify an SAE viscosity grade requirement for some applications. Gear oils intended to be used only in such applications are not required to contain an SAE viscosity grade on their labels.

3.13.3.2. Service Category. – The label on each container of gear oil shall contain the service category, or categories, in letters not less than 3.18 mm ($^1/_8$ in) in height, as defined by the latest version of SAE J308, "Axle and Manual Transmission Lubricants."

(Added 2004)

3.14. Automatic Transmission Fluid.

3.14.1. Labeling. – The label on a container of automatic transmission fluid shall not contain any information that is false or misleading. In addition, each container of automatic transmission fluid shall be labeled with the following:

(a) the brand name;

(b) the name and place of business of the manufacturer, packer, seller, or distributor;

(c) the words "Automatic Transmission Fluid";

(d) the duty type of classification; and

(e) an accurate statement of the quantity of the contents in terms of liquid measure.

3.14.2. Documentation of Claims Made Upon Product Label. – Any manufacturer or packer of any product subject to this article and sold in this state shall provide, upon request of duly authorized representatives of the Director, documentation of any claim made upon their product label.

(Added 2004)

3.15. Biodiesel and Biodiesel Blends.

3.15.1. Identification of Product. – Biodiesel shall be identified by the term "biodiesel" with the designation "B100." Biodiesel blends shall be identified by the term "Biodiesel Blend."

3.15.2. Labeling of Retail Dispensers.

3.15.2.1. Labeling of Grade Required. – Biodiesel shall be identified by the grades S15 or S500. Biodiesel blends shall be identified by the grades No. 1-D, No. 2-D, or No. 4-D.

3.15.2.2. EPA Labeling Requirements also Apply. – Retailers and wholesale purchaser-consumers of biodiesel blends shall comply with EPA pump labeling requirements for sulfur under 40 CFR § 80.570.

3.15.2.3. Automotive Fuel Rating. – Biodiesel and biodiesel blends shall be labeled with its automotive fuel rating in accordance with 16 CFR Part 306.

3.15.2.4. Biodiesel Blends. – When biodiesel blends greater than 20 % by volume are offered by sale, each side of the dispenser where fuel can be delivered shall have a label conspicuously placed that states "Consult Vehicle Manufacturer Fuel Recommendations."

The lettering of this legend shall not be less that 6 mm (¼ in) in height by 0.8 mm (1/32 in) stroke; block style letters and the color shall be in definite contrast to the background color to which it is applied.

3.15.3. Documentation for Dispenser Labeling Purposes. – The retailer shall be provided, at the time of delivery of the fuel, a declaration of the volume percent biodiesel on an invoice, bill of lading, shipping paper, or other document. This documentation is for dispenser labeling purposes only; it is the responsibility of any potential blender to determine the amount of biodiesel in the diesel fuel prior to blending.

3.15.4. Exemption. – Biodiesel blends that contain less than or equal to 5 % biodiesel by volume are exempted from the requirements of Sections 3.15.1. Identification of Product, 3.15.2. Labeling of Retail Dispensers, and 3.15.3. Documentation for Dispenser Labeling Purposes when it is sold as "diesel fuel" as required in Section 3.3. Diesel Fuel.

(Added 2005) (Amended 2008)

Section 4. Retail Storage Tanks and Dispenser Filters

4.1. Water in Gasoline-Alcohol Blends, Biodiesel Blends, E85 Fuel Ethanol, Aviation Gasoline, and Aviation Turbine Fuel. – No water phase greater than 6 mm (¼ in) as determined by an appropriate detection paste or other acceptable means, is allowed to accumulate in any tank utilized in the storage of gasoline-alcohol blend, biodiesel, biodiesel blends, E85 fuel ethanol, aviation gasoline, and aviation turbine fuel.

(Amended 2008 and 2012)

4.2. Water in Gasoline, Diesel, Gasoline-Ether, and Other Fuels. – Water shall not exceed 25 mm (1 in) in depth when measured with water indicating paste or other acceptable means in any tank utilized in the storage of diesel, gasoline, gasoline-ether blends, and kerosene sold at retail except as required in Section 4.1. Water in Gasoline-Alcohol Blends, Biodiesel Blends, E85 Fuel Ethanol, Aviation Gasoline, and Aviation Turbine Fuel.

(Amended 2008 and 2012)

4.3. Dispenser Filters.

4.3.1. Engine Fuel Dispensers.

(a) All gasoline, gasoline-alcohol blends, gasoline-ether blends, E85 fuel ethanol and M85 methanol dispensers shall have a 10 micron or smaller nominal pore-sized filter.

(b) All biodiesel, biodiesel blends, diesel, and kerosene dispensers shall have a 30 micron or smaller nominal pore-sized filter.

4.3.2. Delivery of Aviation Fuel and Gasoline.

(a) Fuel delivery of aviation turbine fuel into aircraft shall be filtered through a fuel filter/separator conforming to API 1581,"Specification and Qualification Procedures for Aviation Jet Fuel Filter/Separators."

(b) Fuel delivery of aviation gasoline into aircraft shall be filtered through a fuel filter/separator conforming to API 1581, "Specification and Qualification Procedures for Aviation Jet Fuel Filter/Separators."

(Added 2008)

4.4. Product Storage Identification.

4.4.1. Fill Connection Labeling. – The fill connection for any fuel product storage tank or vessel supplying engine-fuel devices shall be permanently, plainly, and visibly marked as to the product contained.

(Amended 2008)

4.4.2. Declaration of Meaning of Color Code. – When the fill connection device is marked by means of a color code, the color code shall be conspicuously displayed at the place of business.

4.5. Volume of Product Information. – Each retail location shall maintain on file a calibration chart or other means of determining the volume of each regulated product in each storage tank and the total capacity of such storage tank(s). This information shall be supplied immediately to the Director.

Section 5. Condemned Product

5.1. Stop-Sale Order at Retail. – A stop-sale order may be issued to retail establishment dealers for fuels failing to meet specifications or when a condition exists that causes product degradation. A release from a stop-sale order will be awarded only after final disposition has been agreed upon by the Director. Confirmation of disposition shall be submitted in writing on form(s) provided by the Director and contain an explanation for the fuel's failure to meet specifications. Upon discovery of fuels failing to meet specifications, meter readings and physical inventory shall be taken and reported in confirmation for disposition. Specific variations or exemptions may be made for fuels designed for special equipment or services and for which it can be demonstrated that the distribution will be restricted to those uses.

5.2. Stop-Sale Order at Terminal or Bulk Plant Facility. – A stop-sale order may be issued when products maintained at terminals or bulk plant facilities fail to meet specifications or when a condition exists that may cause product degradation. The terminal or bulk storage plant shall immediately notify all customers that received those product(s) and make any arrangements necessary to replace or adjust to specifications those product(s). A release from a stop-sale order will be awarded only after final disposition has been agreed upon by the Director. Confirmation of disposition of products shall be made available in writing to the Director. Specific variations or exemptions may be made for fuels used for blending purposes or designed for special equipment or services and for which it can be demonstrated that the distribution will be restricted to those uses.

Section 6. Product Registration

6.1. Engine Fuels Designed for Special Use. – All engine fuels designed for special use that do not meet ASTM specifications or standards addressed in Section 2. Standard Fuel Specifications shall be registered with the Director on forms prescribed by the Director 30 days prior to when the registrant wishes to engage in sales. The registration form shall include all of the following information:

6.1.1. Identity. – Business name and address(es).

6.1.2. Address. – Mailing address, if different than business address.

6.1.3. Business Type. – Type of ownership of the distributor or retail dealer, such as an individual, partnership, association, trust, corporation, or any other legal entity or combination thereof.

6.1.4. Signature. – An authorized signature, title, and date for each registration.

6.1.5. Product Description. – Product brand name and product description.

6.1.6. Product Specification. – A product specification sheet shall be attached.

6.2. Renewal. – Registration is subject to annual renewal.

6.3. Re-registration. – Re-registration is required 30 days prior to any changes in Section 6.1. Engine Fuels Designed for Special Use.

6.4. Authority to Deny Registration. – The Director may decline to register any product that actually or by implication would deceive or tend to deceive a purchaser as to the identity or the quality of the engine fuel.

6.5. Transferability. – The registration is not transferable.

Section 7. Test Methods and Reproducibility Limits

7.1. ASTM Standard Test Methods. – ASTM Standard Test Methods referenced for use within the applicable Standard Specification shall be used to determine the specification values for enforcement purposes.

7.1.1. Premium Diesel. – The following test methods shall be used to determine compliance with the premium diesel parameters:

(a) **Cetane Number.** – ASTM D613, "Standard Test Method for Cetane Number of Diesel Fuel Oil";

(b) **Low Temperature Operability.** – ASTM D4539, "Standard Test Method for Filterability of Diesel Fuels by Low-Temperature Flow Test (LTFT) or ASTM D2500, "Standard Test Method for Cloud Point of Petroleum Products" (according to marketing claim);

(c) **Thermal Stability.** – ASTM D6468, "Standard Test Method for High Temperature Stability of Middle Distillate Fuels" (180 min, 150 °C); and

(d) **Lubricity.** – ASTM D6079, "Standard Test Method for Evaluating Lubricity of Diesel Fuels by the High Frequency Reciprocating Rig (HFRR)."

(Amended 2003)

7.2. Reproducibility Limits.

7.2.1. AKI Limits. – When determining the antiknock index (AKI) acceptance or rejection of a gasoline sample, the AKI reproducibility limits as outlined in the latest version of ASTM D4814, "Standard Specification for Automotive Spark-Ignition Engine Fuel, Appendix X1 shall be acknowledged for enforcement purposes.

7.2.2. Reproducibility. – The reproducibility limits of the standard test method used for each test performed shall be acknowledged for enforcement purposes, except as indicated in Section 2.2.1. Premium Diesel Fuel and Section 7.2.1. AKI Limits. No allowance shall be made for the precision of the test methods for aviation gasoline or aviation turbine fuels.

(Amended 2008)

7.2.3. SAE Viscosity Grades for Engine Oils. – All values are critical specifications as defined in the latest version of ASTM D3244, "Standard Practice for Utilization of Test Data to Determine Conformance with Specifications." The product shall be considered to be in conformance if the Assigned Test Value (ATV) is within the specification.

(Added 2008)

7.2.4. Dispute Resolution. – In the event of a dispute over a reported test value, the guidelines presented in the latest version of ASTM D3244, "Standard Practice for Utilization of Test Data to Determine Conformance with Specifications," shall be used to determine the acceptance or rejection of the sample.

7.2.5. Additional Enforcement Action. – The Director may initiate enforcement action in the event that, based upon a statistically significant number of samples, the average test result for products sampled from a particular person is greater than the legal maximum or less than the legal minimum limits (specification value), posted values, certified values, or registered values.

(Added 2008)

V. Examination Procedure for Price Verification

as adopted by
The National Conference on Weights and Measures*

A. Background

The NCWM established the Price Verification Working Group in 1993 to respond to public concern about price accuracy in retail stores. More than 500 retailers, consumer representatives, and state and local weights and measures officials participated in the development of the procedure. It was adopted by the NCWM at the 80th Annual Meeting in 1995.

The procedure applies to all retail stores, including food, hardware, general merchandise, drug, automotive supply, convenience, and club or other stores. Model inspection reports are included to promote the collection of uniform data. The model reports and uniform procedures will serve as the foundation for the collection and summarization of price accuracy data on a national basis. This information may be used to provide reliable information on price accuracy with a national perspective. The procedure provides administrators with the tools, guidance, and background information, as well as uniform test procedures and enforcement practices, to enhance the economic well-being of consumers and retail businesses in their jurisdiction. By implementing this program in cooperation with industry, officials will help to restore and maintain consumer confidence in retail pricing practices and technologies, such as scanners, and provide economic benefits for consumers and the business community.

B. Status of Promulgation

The Examination Procedure for Price Verification was recommended for adoption by the Conference in 1995. The table beginning on page 10 shows the status of adoption of the procedure.

**The National Conference on Weights and Measures (NCWM) is supported by the National Institute of Standards and Technology (NIST) in partial implementation of its statutory responsibility for "cooperation with the states in securing uniformity in weights and measures laws and methods of inspection."*

THIS PAGE INTENTIONALLY LEFT BLANK

Examination Procedure for Price Verification

Table of Contents

Examination Procedure for Price Verification

Section 1. Scope

These procedures may be used to conduct price verification inspections in any type of store, including those that use Universal Product Code (UPC) scanners and price-look-up codes at the check-out counter as a means for pricing. Procedures are included for test purchases and verifying manual entries. The purpose of the procedure is to ensure that consumers are charged the correct price for the items they purchase. The "randomized" and "stratified" sampling procedures are intended for use in routine inspections to determine how well a store is maintaining price accuracy. Nothing in this procedure should be construed or interpreted to redefine any state or local law or limit any jurisdiction from enforcing any law, regulation, or procedures that relates to the accuracy of advertisements of retail prices, or any other legal requirement.

Section 2. Definitions

2.1. Area. – "Entire store," a "department," "grouping of shelves or displays," or other "section" of a store as defined by the inspector from which samples are selected for verification. "Non-public" areas of a store are not included (e.g., the area in a pharmacy where controlled drugs are kept or product store rooms).

2.2. Cents-off Representation. – Any printed matter consisting of the words "cents off" or words of similar import placed upon any item or on a label affixed or adjacent to an item, stating or representing by implication that it is offered for sale at a lower price than the ordinary and customary retail selling price (e.g., 15 % off, bonus offers, 2 for 1, or 1-cent sales, etc.).

2.3. Direct-Store-Delivery (DSD) Item. – An item delivered to a store, and usually priced, by route salespeople (e.g., milk, beer, soft drinks, bread, and snack foods).

2.4. Displays.

(a) **Aisle Stacks or End-of-Aisle Displays.** – Displays located in freestanding units or attached at the end of or adjacent to a tier of shelves.

(b) **Tie-in Displays.** – Displays of related products at secondary locations in a store (e.g., barbecue sauce on shelves in an aisle that may also be simultaneously displayed in the meat department of a food store).

(c) **Multiple Displays.** – Displays of the same product at several locations in a store.

2.5. Hand-held Scanning Device. – A portable device that scans UPC codes and also allows for the comparison of the price displayed on a shelf, item, or otherwise advertised to the price for the item in the point-of-sale database.

NOTE: These devices either retain a "batch" file of entered prices and identities for later comparison to the database or operate "on-line" via FM radio to the database. When used for price verification, they shall only be used with the active point-of-sale database. If you use a hand-held scanner, verify all price discrepancies by scanning the item at a check-out register and request a printed receipt to document the price that consumers would be charged.

2.6. Enforcement Levels.

NOTE: These recommendations are not intended to modify the enforcement policy of any jurisdiction unless they are adopted by the jurisdiction.

(a) **Lower levels of enforcement actions.** – Includes increased inspection frequency, stop-sale or correction orders, warning letters, and other notifications of noncompliance.

(b) **Higher levels of enforcement actions.** – Includes issuance of citations, administrative hearings, civil penalties, or prosecution under criminal statutes.

2.7. Inspection Types.

(a) **Automated Inspection.** – Inspections that are conducted using a hand-held scanning device.

(b) **Manual Inspection.** – Removing items from displays and taking them to a check-out terminal to verify the price (e.g., select the items and either (1) take them to a check-out terminal for scanning or (2) record the product identity, UPC number, and shelf price for each package on an inspection report) and then manually entering the UPC numbers in the register. The manual entries may be made by the official or by a store employee.

2.8. Inspection Frequency.

These recommendations do not modify the inspection policy of any jurisdiction unless adopted by the jurisdiction.

Inspection Control. *– After a program has been in place for a period of time and a database is established, procedures can be developed to randomly select stores for inspection, or to focus inspections on stores with low levels of compliance.*

(a) **Normal Inspection Frequency.** – An inspection made at the customary time interval used by an enforcement agency. Inspections may be conducted during normal business hours. Stores under this normal frequency should be inspected semi-annually or annually.

(b) **Increased Inspection Frequency.** – An inspection made more often than with the customary time interval, usually as a follow-up on prior violations. Inspections may be conducted during the normal business hours. Stores under this increased frequency should be inspected on a quarterly, bi-monthly, or more frequent basis.

(c) **Term of Increased Inspection Frequency.** – A store placed on an increased inspection frequency shall remain at that frequency until there are two consecutive inspections with an accuracy of 98 % or higher.

(d) **Special Inspection.** – An inspection that is conducted as a follow-up to a prior inspection or to investigate a complaint.

2.9. Inspection Lot. – A group of items available for testing in an "area" or "areas." (See 2.1. "Area.")

2.10. Merchandise Group. – A group of products identified under a common heading for inspection purposes only (e.g., "advertised sale" items, "end-of-aisle" items, "direct delivery" items, "cents-off" items, or all the items in the "men's" department in a department store).

2.11. Not-on-File Item. – Items not found in the point-of-sale database. When found, another item is selected at random (e.g., an item on either side of the one that was not on file) to replace the item in the sample. A "not-on-file" item is not an error unless you determine that the price "charged" for the item is incorrect by conducting a test purchase or by asking the check-out clerk to determine the price by using the store's written or stated policy or procedures. If the price is found to be inconsistent, the error is included in the total.

2.12. Notification of Noncompliance. – Any written notice given to a store describing the violations of the law that were found.

2.13. Price Look-Up Code (PLU). – A pricing system where numbers are assigned to items or commodities, and the price is stored in a database for recall when the numbers are manually entered. PLU codes are used with scales, cash registers, and point-of-sale systems.

2.14. Prices. – These definitions do not amend or effect the provisions of any law, regulation, or other test procedure.

(a) **Misrepresented Price.** – The price charged differs from the price at which the item is offered, exposed, or advertised for sale, or that the price is different from the price on the item, shelf label, or sign.

(b) **Price Charged.** – The price charged for an item and either displayed on the automated device or on the receipt issued by the device, whether the item is scanned or actually purchased, the device is computing or recording while in a training or inspection mode, or by using the hand-held device tied to the point-of-sale database.

(c) **Overcharge.** – The price charged for an item is more than the lowest advertised, quoted, posted, or marked price.

(d) **Undercharge.** – The price charged for an item is less than the lowest advertised, quoted, posted, or marked price.

(e) **Intentional Undercharge.** – Undercharges are not counted as errors if the store provides, at the time of inspection, information that confirms that the price error was intentional (e.g., an undercharge that occurs when a store lowers a price in a database before it changes shelf tags or signs in anticipation of selling the item at a lower price, or when a store increases the price or advertised price of an item, and then increases the price in the database, or when a discounted price is rounded to a lower value).

2.15. Pricing Coordinator. – The individual designated by the store to control and maintain "pricing integrity" in the store, although the title will differ among retailers.

2.16. Pricing Integrity. – Ensuring that the computer price file and/or the price charged to consumers at a cash register is the same price that is marked on the product, in an advertisement, and/or the shelf tag.

2.17. Sample. – The number of items selected for testing from the inspection lot.

2.18. Scanner. – An electronic system that employs a laser bar code reader to retrieve product identity, price, and other information stored in computer memory.

2.19. Stock-Keeping Unit (SKU). – A system of product identity and pricing similar to PLUs.

2.20. Store-Coded Item. – The application of UPC codes to items in the store. Scales in the meat, deli, and other departments generate UPC labels that include identity and price information that can be read by point-of-sale scanners.

2.21. Stop-Sale Order. – An official document placing a package or an amount of any commodity off-sale, that is offered or exposed for sale in violation of the law.

2.22. Ticketed Merchandise. – Items from which the price must be read from a ticket (or price sticker) and manually keyed into a register.

2.23. Universal Product Code (UPC). – A unique symbol that consists of a machine readable code and human-readable numbers. UPCs are printed on package labels or are applied with tags or labels. UPC codes may be printed for random weight packages by price computing scales. UPC symbols must meet the standards established by the GS1 US (formerly the Uniform Code Council [UCC]) in order for them to "scan" accurately. The size and clarity of the print and clear area surrounding the symbol are just a few of the factors that affect accuracy. The GS1 US issues codes and answers technical questions. For more information, contact GS1 US, at 7887 Washington Village Drive, Suite 300, Dayton, OH 45459, telephone: (937) 435-3870, FAX (937) 435-7317, or e-mail **info@gs1us.org**. You can visit them on the web at **www.gs1us.org**.

Section 3. Test Notes

3.1. Safety and Health. – Practice safe work habits to avoid personal injuries or property damage. Be aware of and follow all safety or sanitation rules at the inspection site. Handle perishable, dairy, or frozen products properly to avoid damage (e.g., avoid defrosting frozen foods or allowing dairy products to warm to room temperature that may result in spoilage).

3.2. Confidentiality of Findings. – Inspection findings should be discussed only with an authorized store representative and released only in accordance with applicable public records laws.

Section 4. Materials and Equipment

The following materials and equipment are recommended for use in conducting the inspections in this procedure:

Inspection report:

- Copy of laws or regulations
- Hand-held counter or Price Verification Tally Sheets
- 1 lb (or 1 kg) test standard
- Merchandise cart (if required and available)

Other equipment and materials provided by the store when available:

- Current newspaper advertisement or store sales brochures
- Hand-held scanning device(s) – Stores are not required to have this equipment or to make it available for your use. However, many stores use this equipment to maintain price integrity and may make it available for your use on request.

Section 5. Pre-Inspection Tasks

Prior to conducting an inspection, it is recommended that you contact the store management, identify yourself, and explain the purpose of your visit. Determine if there are any health, sanitation, or safety rules. If requested, provide information on the law or the inspection procedure.

NOTE: *When verifying manual price entries or conducting test purchases, store management is typically not notified of the test until the items have been totaled and the transaction completed.*

(a) Notify store representatives that they are invited to participate in the inspection.

(b) If the store makes a hand-held scanning device available for use, request instructions on how to operate it properly. It is acceptable for the "pricing coordinator" to operate the scanning device and participate in the inspection.

(c) If you use the manual inspection procedure, advise the store representative that you will return the merchandise to its display location unless the store representative wants to restock the items, which is acceptable. Determine which check-out location to use. Arrange to have the register set so that the items you verify are not included in sales records.

(d) Conduct inspections in a manner that does not disrupt normal business activities.

Section 6. Inspection

Perform the following inspections:

6.1. Position of Equipment. – Determine if customer indications on point-of-sale systems meet NIST Handbook 44, General Code, User Requirement, 3.3. Position of Equipment. A device equipped with a primary indicating element and used in direct sales shall be so positioned that its indications may be accurately read and the weighing and measuring operation may be observed from some "reasonable" customer position.

NIST Handbook 44 defines "point-of-sale system" as an assembly of elements including a weighing element, indicating element, and a recording element (and may be equipped with a scanner) used to complete a direct sale transaction.

NOTE: *The importance of consumer* access *to the cash register display of product information and price cannot be overstated. If consumers cannot verify prices as the items are being scanned, they must wait until the transaction is completed (i.e., they must pay by cash, check, or credit card) before they receive the receipt and can confirm the prices charged for the items.*

6.2. Other.

(a) If you use a cash register, verify the accuracy and legibility of information provided on register's receipts.

(b) Conduct inspections to enforce local requirements if your jurisdiction has specific laws or regulations relating to price marking, shelf labels, or unit pricing.

Section 7. Test Procedures

These procedures shall be used to conduct inspections in any type of store, whether the store uses scanners or automated price look-up registers, or where a clerk manually enters the prices.

7.1. Application of Sampling Plans.

(a) For normal or increased frequency inspections, follow the procedures referred to in Columns 1, 2, and 3 in Table 1. Samples, Sample Collection, and Accuracy Requirements.

(b) For special inspections, use the test procedures in Section 7.2. Table 1. Samples, Sample Collection, and Accuracy Requirements or 7.4. Procedure for Test Purchases and for Verifying Manually Entered Prices.

7.2. Table 1. Samples, Sample Collection, and Accuracy Requirements.

7.2.1. How to Use the Table:

(a) Look in Column 1 for the type of store you are inspecting; select the appropriate sample size from Column 2; then refer to Column 3 for the type of sample collection plan to use.

(b) Follow the single-stage or two-stage sampling plans to conduct the inspection and collect the samples using either the "randomized" or "stratified" sample collection procedures described in Section 7.3. Sample Collection Procedures or the procedure in Section 7.4. Procedure for Test Purchases and for Verifying Manually Entered Prices.

(c) Apply the accuracy requirements for the appropriate sample size in Column 4.

7.2.2. Samples. – Refer to Column 2 in Table 1. Samples, Sample Collection, and Accuracy Requirements to determine how many items to select for the store type and whether to use the single-stage or two-stage sampling plan. You may use either:

(a) **Single-Stage Sample.** – A single-stage sample is typically used for, but is not limited to, stores where a hand-held scanner device is available for the inspection; or

(b) **Two-Stage Sample.** – A two-stage sample saves time. If the sample (usually one-half the total sample size) taken in the first-stage meets the accuracy requirements specified in Column 4 in Table 1. Samples, Sample Collection, and Accuracy Requirements, the inspection is complete. However, if the errors in the first-stage sample fall within the limits set in Column 4, the second-stage of the sample is taken.

7.3. Sample Collection Procedures (for use with either manual or automated inspection procedures). – These sample collection procedures may be used to conduct either manual or automated inspections with a single-stage or two-stage sample. That is, you can either use a hand-held scanning device to verify the price of an item (automated), or you can remove the items from display and take them to a check-out location to verify the price of the item (manual) regardless of which sample collection procedure is used. No sample collection procedure is ideal for all retail store arrangements. You can modify the procedure to fit each store, but you should adhere to the sample size and sample collection procedures described in Table 1. Samples, Sample Collection, and Accuracy Requirements. When using any of the procedures, test the store as a whole unit by taking samples from all parts of the store, or divide the store into "areas" and select samples from several "areas" (e.g., at least 10 areas, or one-third or one-half of the "areas").

<table>
<tr><th colspan="4">Table 1.
Samples, Sample Collection, and Accuracy Requirements</th></tr>
<tr><th>Column 1.
Type of Store</th><th>Column 2.
Samples</th><th>Column 3.
Sample Collection Procedures</th><th>Column 4.
Accuracy Requirements
(See Section 10)</th></tr>
<tr><td>Convenience or Any Other Small Retail Store

NOTE: For this procedure, a small store is typically one with three or fewer check-out registers.</td><td>Two-Stage Sample:

First Stage = 25 items

Second Stage = 25 items or more

Total = 50 items or more

or

Single-Stage Sample:

50 items or more</td><td rowspan="3">Use the Randomized Sample Collection in 7.3.1 or the Stratified Sample Collection in 7.3.2.

and

Use Manual or Automated Inspection Procedures

NOTE: Test the store as a whole unit by taking samples from all "areas" of the store,

or

divide the store into "areas" and select samples from several "areas" (e.g., at least 10 or one-third of the "areas")</td><td>If 1 error is found in the 25-item sample, test an additional 25 items.

If more than 1 error is found in the 50-item sample, the store fails.

NOTE: If more than 1 error is found in the first 25 items, the store fails.</td></tr>
<tr><td>All Other Retail Stores</td><td>Two-Stage Sample:

First Stage = 50 items

Second Stage = 50 items or more

Total = 100 items or more</td><td>If 1 error is found in the 50-item sample, the store passes.

If 2 errors are found in the 50-item sample, test an additional 50 items.

If more than 2 errors are found in the 100 item sample, the store fails.

NOTE: If more than 2 errors are found in either stage, the store fails.</td></tr>
<tr><td></td><td>Single-Stage Sample:

100 or more items</td><td>If more than 2 errors are found in the 100-item sample the store fails; or

If more than 100 items are sampled, the error rate shall not exceed 2 %.</td></tr>
</table>

NOTE 1: *These sampling procedures allow flexibility in sample collection for use in any type or size of store. You can take several different approaches and select a number of "areas" to sample using the sample sizes in Table 1. For example, to perform a 100-item inspection in a department store with 20 "areas," you can either verify 5 items in an "area," 10 items in each of 10 "areas," or 20 items from each of 5 "areas."*

NOTE 2: *The sample sizes used for routine inspections in this procedure should not be used to estimate the overall accuracy of prices in a store.*

NOTE 3: *In some stores, price reductions are not programmed into the point-of-sale system. Instead, discounts are manually entered by a sales clerk; however, the sales clerks should have a means of identifying a sale item. When conducting normal inspections, verify the price of the sale items by allowing the sales clerk to determine the price of the item using the store's customary procedures. This will ensure that the customer receives the correct price regardless of the location where the check-out occurs.*

7.3.1. Randomized Sample Collection. – In "randomized" sample collection, all items in an "area" have an equal chance of being included in the sample. This test procedure has several benefits, including: (1) having more effective coverage and being simpler to conduct because you select items by count following a systematic pattern throughout the store, and (2) ensuring that a wider range of items are verified, which increases scrutiny; therefore, there is greater confidence in the results. With most samples, several items will be verified in each "area" of the store. Since store sizes differ, this number will vary, but samples should be taken from a wide variety of items (and merchandise groups) from locations throughout the store or "area."

The steps of the randomized sampling collection procedure are as follows:

(a) Count the number of "areas" in the store which have products to be verified:

(1) Stand-alone counters and displays or whole departments (e.g., bakery or seafood, or "men's clothing" or "sporting goods" department, etc.) are considered and counted as individual "areas" to be sampled.

(2) End of aisle displays may be considered as a single, distinct "area" and either verified separately or included as part of one side of an aisle.

(b) The sample size (e.g., 100 items) is divided by the number of "areas" to determine the number of items to be sampled from each "area." Depending on the number of areas in the store, you may calculate a fractional number of items per area. In this case, round off the sample size and select one or two additional items from an "area" to complete the full sample size of 100 items.

7.3.1.1. Example 1. Illustrations of the Randomized Sampling Procedure.

(a) Figure 1 illustrates how the randomized sampling procedures are used in a food store. This example is based on a 100-item sample. To simplify the selection process, simply divide the store into 4 major "areas" and select samples as follows:

Examples:

- Select 5 items from all of the shelves and displays in the produce section which are grouped as a single "area,"
- Select 85 items by choosing 5 items from either side of several of the 13 aisles (e.g., there are 26 rows of shelves from which samples may be selected. To select 85 items, select 5 items from 17 of the 26 rows of shelves).
- Select 5 items from the counters along the back of the store, and
- Select 5 items from the deli-bakery and the cash register areas which are grouped as a single "area."

(b) Figures 2 and 3 illustrate how the randomized sampling procedures may be used in any store. The examples are based on a 100-item sample for stores that have a total of 30 "areas" to sample. The procedure allows the flexibility needed to adjust the sample to fit the store layout. To simplify the selection process, the stand-alone displays may be grouped together as an "area" to be sampled.

The following breakdown of "areas" is illustrated in Figure 2; the same approach is used in Figure 3. Figure 4 illustrates an example of sampling 100 items by selecting 20 items from 5 different areas in a department store.

1 - All shelves and displays in the produce section are grouped as a single "area."
28 - The 13 aisles (26 rows of shelves), the counters along the back of the store, and the cash register areas are counted as "areas."
1 - The "end-of-aisle" displays at the front and back of the store are grouped as a single "area."
30 - Total "areas"

(1) To select samples from the entire store, divide 100 by 30 to calculate how many "samples" to take from each "area." In this example, 100 ÷ 30 = 3.3 items per area. Rounding down to 3 items, take a total of 90 samples from the different "areas," then select an additional one (1) item from each of 10 "areas" to obtain a sample of 100 items.

(2) If you round up to 4 items per area, you take a total of 120 samples, or

(3) You may select 10 items from 10 "areas."

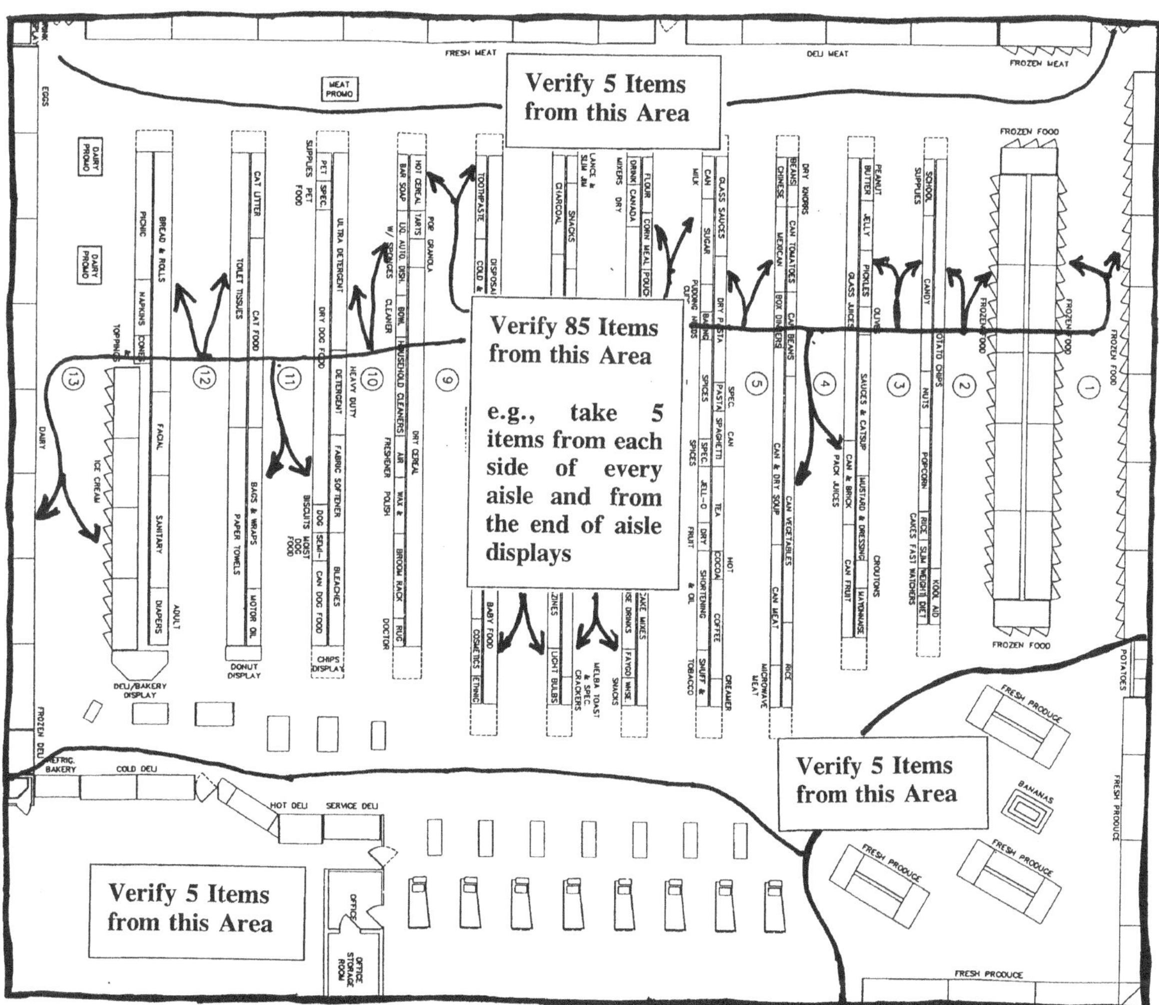

Figure 1. Illustration of the Randomized Sampling Procedure

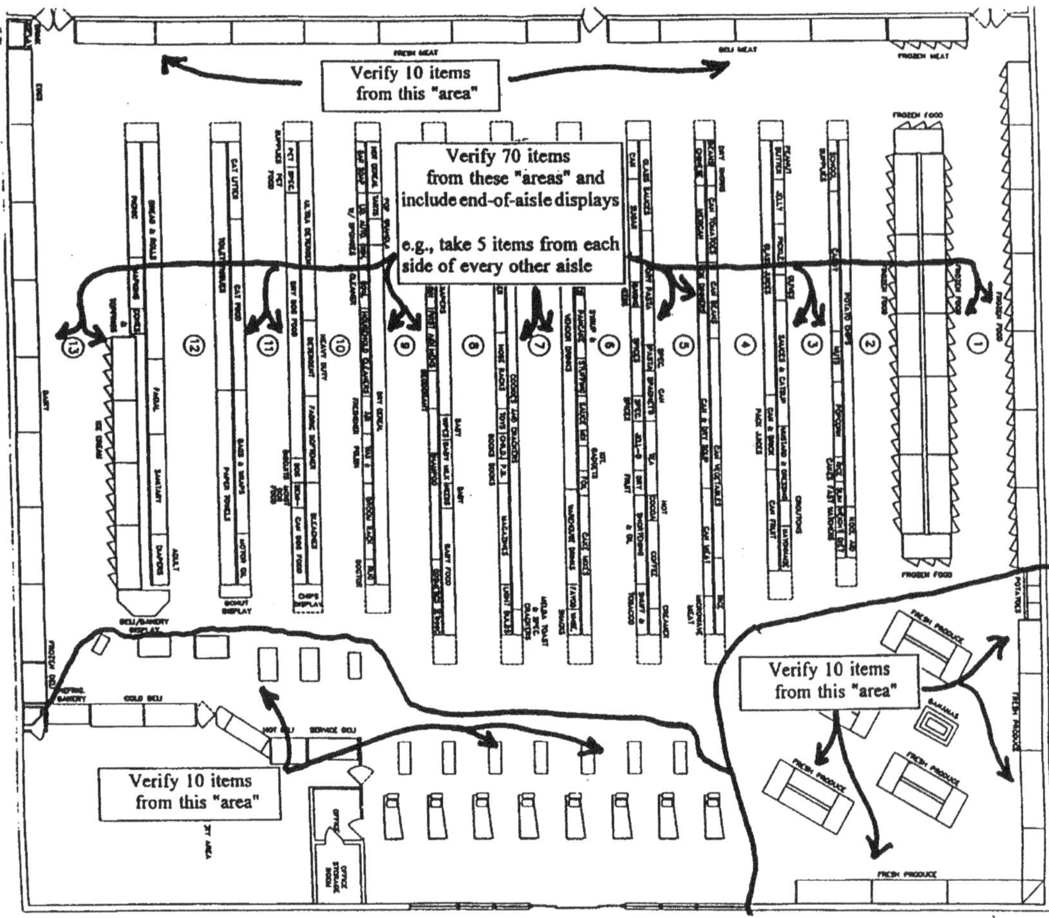

Figure 2. Illustration of the Randomized Sampling Procedure

(c) Start in any "area" in the store at any shelf, rack, or display (top, bottom, front, back; anywhere on a circular rack or display). Begin with the first, second, or third item and count either 5, 10, or 15 items along the shelf (varying the number of items counted depending on how many items are available on the shelf) or along the aisle. Select the 5th, 10th, or 15th item as appropriate (See Figures 5, 6, and 7). Only select one item from each brand or product (if they are the same price) from a display that has two or more items of the same product size and price displayed side by side. You can change the number of items you count off as often as necessary during the inspection.

(d) Either verify the price with a hand-held scanning device or take the item (along with the other items you select) to the check-out location to verify the price, keeping count of the items using a hand counter or tally sheet. If the price of an item is incorrect, record the item's name, description, and price along with other information (e.g., whether the product is on sale, aisle location so you can easily find the items again to verify the error, etc.).

(e) From the first item sampled, move down (or up) one shelf to the item most directly below (or above) and count 5, 10, or 15 items in the same direction and sample the 5th, 10th, or 15th items, as appropriate. After the number of items to be verified in each "area" have been selected, go to the next "area" and start on the next shelf (either down or up) from where the previous sample was selected, count 5, 10, or 15 items and select the appropriate item using the count system until the required number of samples is selected. If you have sampled an item on the bottom (or top) shelf and have more items to test in the "area," simply go up (or down) one shelf. This will create a "zigzag" trail up and down the display.

NOTE: *Randomness can be increased by starting on different shelves or at the midpoint or rear of an aisle during an inspection, or by starting at different locations in a store on subsequent inspections. Always start at a different location on subsequent inspections of a store. To maintain "randomness," do not search for obvious pricing errors. If you see pricing errors, have them corrected. The sample should not include more than one of the same item from the same display. If an item is out of stock, select the next item.*

(f) This procedure is repeated for all "areas" until you complete the sample. (See following Notes)

NOTE 1: *Include at least 5 to 10 Price Look Up (PLU) and store-coded items in the samples. In food stores, these items do not usually have to be removed from the produce, bulk foods section, or deli display for use in this procedure. You can use a hand-held scanner or record the identity and item price designated at the product sales display of the items from the different department (produce, bakery, deli), if available, for price comparison through either the PLU programmed in the department's scale or at the point-of-sale system. Have the PLU entered in the scale (See Note 2) or point-of-sale system (or have "store-coded" items scanned) and record the price, comparing it with the displayed sale price. Record any errors (See Note 3). When checking "store-coded" items from the meat or other departments, remember a "UPC symbol" on a random weight label is read by a scanner to obtain the total price and identity. The price is not stored in the point-of-sale database, but in the memory of the prepackaging scale.*

NOTE 2*: Some scales or point-of-sale systems do not display or record the unit price associated with the PLU unless a weight is on the scale. For this type of device, a one pound standard (or 1 kg) is placed on the scale load-receiving element. Some systems automatically deduct tare, so check to make sure that this does not affect the price indication.*

NOTE 3: *When you manually enter PLU codes and find errors, reenter the PLU number to ensure that the error was not caused by a keying mistake and that the item was identified accurately.*

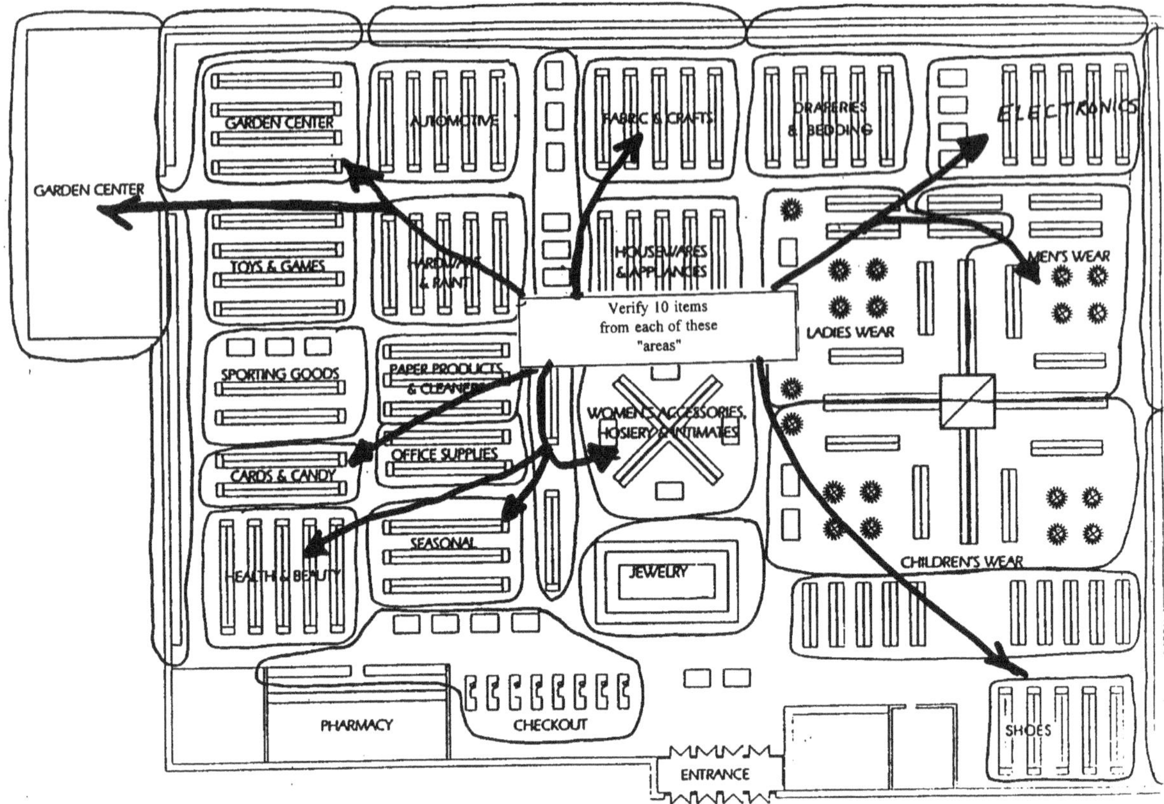

Figure 3. Illustration of the Randomized Sampling Procedure

7.3.2. Stratified Sample Collection. – Stratified sample collection (i.e., selecting samples from specific merchandise groups) of items on sale, specials, seasonal items, or items on end-of-aisle displays) is typically used (e.g., if a store has failed an inspection based on the randomized sample collection procedures) to focus on specific merchandise groups that appear to have more errors than others (e.g., you find that many of the errors found in the randomized sample were in "advertised specials" or with "discontinued items"). You can also combine sample collection procedures by using a "randomized/stratified" approach. The "stratified" approach may be used the first time you inspect a store, in stores that have just implemented scanning, in stores that have high error rates on particular groups of items in past inspections, or in responding to consumer complaints involving a particular group of items.

For stratified sample collection, items are randomly selected from different "merchandise groups" in a store. They are tested in the first stage of the two-stage manual sampling plan to determine if (1) any group has more errors than any other and (2) the sample taken in the first stage meets accuracy requirements. This method should be modified depending on the marketing practices of the store in which it is used (e.g., if you are in a department store, there may be fewer groups to sample from, or the list provided below may not include the types of groups typically encountered in a hardware superstore). The next example shows how to conduct a stratified sample and how it is used, but it should not be the sole basis for sample collection because a specific list of items does not look at the store as a whole. Focusing on specific merchandise groups takes time, but this may be necessary when investigating a complaint or following up on a prior noncompliance. Select only one item from each brand or product from a display that has two or more items of the same product, size, and price displayed side by side if they are the same price.

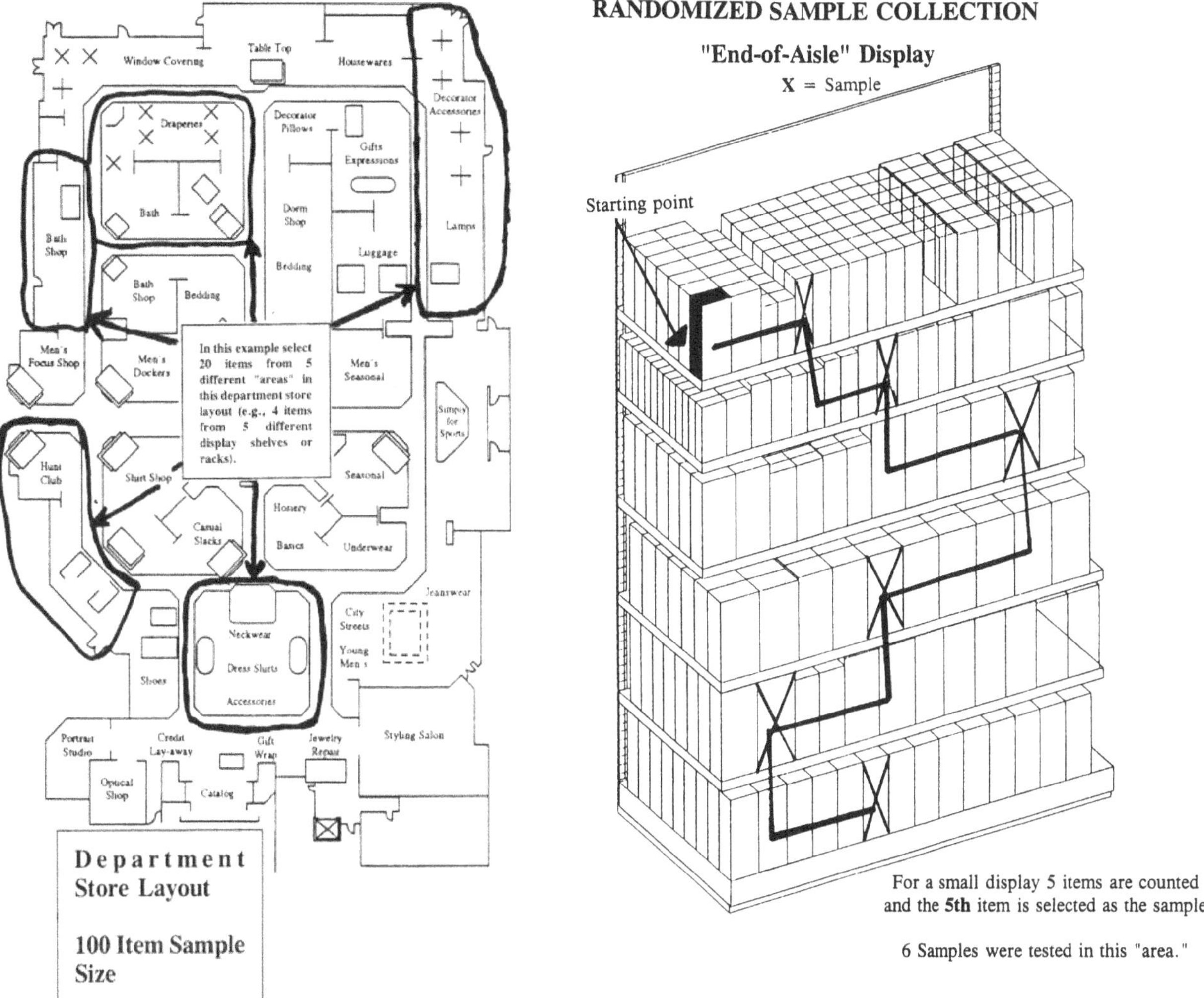

Figure 4. Stratified Sample Collection

Figure 5. Randomized Sample Collection

Sample Size. – In this example, a large food store is inspected using a two-stage sampling plan (50 items/100 total items). The inspection begins with an initial sample of 50 items (see Column 1. Type of Store for All Other Retail Stores and Column 2. Sample Sizes in Table 1).

Stratified Sample Collection. – Select 50 items from the merchandise groups listed below (provided as examples only; stores may have other groups that should be included). This procedure allows you to focus on specific merchandise groups to determine if errors are indeed occurring in groups where they are thought to occur most frequently (e.g., sale and direct delivery items).

7.3.2.1. Example 2. Two-Stage Manual Inspection using the Stratified Sampling Procedure.

If there is an insufficient number of items in any merchandise group, or if the group of items is not available, increase the number of "randomized" items selected from the overall inspection lot to obtain a total of 50 items. As marketing practices evolve, these groups may change as well. You may substitute "other" or new merchandise groups for any of those listed below (e.g., you may have identified errors in the "health and beauty aids" section or on "manager specials" during a previous inspection, so samples from these groups may be substituted for any of the groups listed below). Model "Price Verification Tally Sheets" in Section 14. Model Forms for Price Verification Inspections are provided for your use with the test procedures to keep track of the number of items selected.

First-Stage: 50 items. – Use the "randomized" sample collection procedures described in 7.3.1. Randomized Sample Collection to select the following items. These sample collection procedures simplify the inspection process and ensure that samples are collected as randomly as possible.

Examples:

- Twenty-five "Regular Priced" items. Select one or two items at random from different shelves in each "area" or limit your sampling to shelves in one-half the "areas" in the store, and

- Twenty-five Items. Select a total of 25 items. Include several items from any of the following merchandise groups:

  - "Direct-Store-Delivery (DSD)" items. If the store allows vendors to price DSD items, include those items in the sample.

  - "End-of-Aisle" or "Tie-In-Display" items. This group can include both regular and sale-priced items.

  - "Advertised Sale" items. Use the store's sales brochure or newspaper advertisements to identify sale items.

  - "Special" items. This includes any item with a reduced price (e.g., items on "special" including "cents-off" or "percentage-off" items, 2-for-the-price-of-1 specials, manager and in-store specials, or discontinued items). Items typically discounted on a percentage basis include a manufacturer's product line, greeting cards, magazines, or books.

  - "PLU" items. This includes both regular and sale priced items offered in the produce, bakery, or bulk food departments and over scales at the direct sale counters. For direct service departments (e.g., produce, deli, specialty meats, etc.), select products at random (include some sale or special prices) and enter the code in the scale [*NOTE 1, page* 210] to verify that the coded price matches the advertised price [*NOTE 2, page* 210].

  - "Store-coded" items. This includes items offered in the produce, bakery, or meat departments that have labels with the UPC symbol generated by scales and printers in the store. For store-coded items, scan the item and determine if the total price and identity on the label are accurately read by the point-of-sale system. When checking "store-coded" items from the meat or other departments, remember that a "UPC symbol" on a random weight label is read by a scanner to obtain the total price and identity. The price is not stored in the point-of-sale database.

  - "Other" items. This category is included to provide flexibility in selecting a sample so that "seasonal" items, or products unique to the store or local market, can be included. Both regular and sale-priced items can be included in this category.

NOTE 1: *Some scales or point-of-sale systems do not display or record the unit price associated with the PLU unless weight is on the scale. For these devices, a 1 lb (or 1 kg) standard is placed on the scale load-receiving element. Some systems automatically deduct tare, so make sure this does not affect the price indication.*

NOTE 2: *When a not-on-file item is found, another item is selected at random to replace it in the sample. A "not-on-file" item is not an error unless you determine (e.g., by conducting a test purchase or by asking the check-out clerk to determine the price of the item using the store's customary procedures) that the price "charged" for the item is incorrect. If the price determined is not correct, the error is included in the total.*

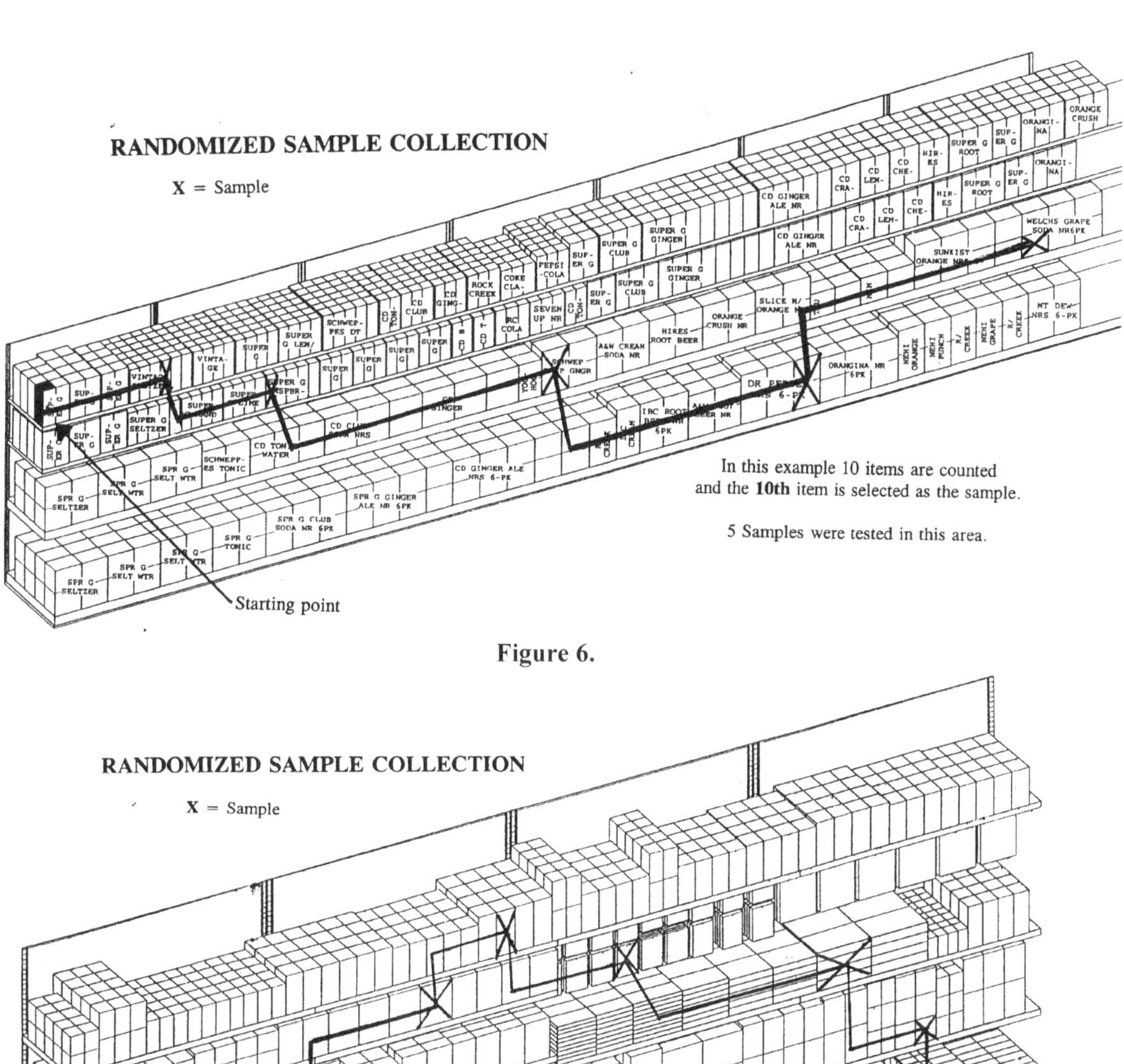

RANDOMIZED SAMPLE COLLECTION
X = Sample
In this example 10 items are counted and the 10th item is selected as the sample.
5 Samples were tested in this area.
Starting point

Figure 6.

Figure 7.

Identify the item on an inspection report (e.g., record a brief description, item number, shelf, or advertised price and aisle location. The aisle location makes it easy to find the product if errors are found and to re-shelve the items). As items are selected, use the "Price Verification Tally Sheet," or other means, to keep track of the number of items collected. (See Section 14. Model Forms for Price Verification Inspections. The "Model Price Verification Reports" in this proposal were developed with the assumption that it is only necessary to record information of items found with price errors, not all items verified. This reduces paperwork and saves time.) Either use a hand-held scanning device or take the items to a cash register, verify the prices by scanning the items or entering a PLU code into the register and printing a receipt. The prices "charged" at the register are then compared to the advertised price of each item. For large or perishable items, record the identity, UPC Code, location, and price and manually enter the UPC number into the register to verify the price. However, this method is subject to recording and key entry errors.

Evaluation of Results on First-Stage.

See Section 9. Evaluation and Inspection Results for guidance on which errors are considered violations: One error in a 50-item sample is permitted. If not more than one error is found and verified, the store passes; if 3 items are found in error in the first 50 items, the store fails and the inspection is complete.

If two errors are found, collect 50 more items using the randomized sampling procedures and verify a total of 100 items. If errors were found in any specific merchandise group (or groups) of items (e.g., direct-store-delivery items, PLU codes, or specials), the additional 50 items should include items from those merchandise groups.

Accuracy.

Refer to Column 4 in Table 1. Samples, Sample Collection, and Accuracy Requirements. The required accuracy is 98 % on the 100-item sample (that is, at most two errors are permitted on a 100-item sample). If more than two errors are found and verified, the store does not meet the accuracy requirement.

NOTE: *The "randomized" and "stratified" sample collection procedures in this section are intended for use in routine inspections to determine how a store is maintaining price accuracy on all of the items it offers for sale. If you use these sampling procedures in routine inspections and uncover a significant number of errors in a particular merchandise group (e.g., a significant number of the pricing errors are found with "advertised sale item" items), a randomized sample can be collected entirely within this specific merchandise group. For example, if the error rate for "advertised specials" is higher than the rate for regular priced items, a more focused inquiry to determine if there is a significant error rate in this merchandise group may be justified. If several "advertised specials" have been the subject of consumer complaints, or if they are repeatedly found to be in error during routine inspections, then a randomized sample can be limited to the "advertised specials" merchandise group. In this case, a randomized sample (e.g., a 50/100 item two-stage approach) is taken from all of the "advertised sale items" offered for sale in the store or in a specific "area." The results of this sample are applicable only to the "advertised specials" group and not to all items in the store.*

7.4. Procedures for Test Purchases, Investigation of Consumer Complaints, and for Verification of Manually Entered Prices.

7.4.1. Procedure. – This procedure may be used to (1) investigate consumer complaints, (2) determine if a store has corrected a pricing error after being notified that an error occurred, or (3) determine if manually keyed-in prices or PLU codes are accurate.

NOTE: *When verifying manual price entries, store management is typically not notified of the test until the items have been totaled and the transaction completed.*

(a) Do not alert the clerk to the fact that the test purchase procedure is being conducted. Do not ask questions concerning any errors that you observe or offer any information if asked the price of an item, in cases where the item price is illegible, or where the item is not on file.

(b) Use the "randomized" sampling procedures to select a sample of 10 to 50 items that includes regular and sale priced items, PLU items, and advertised specials from various "areas." It is acceptable to purchase only one or just a few items if you are investigating a complaint on a specific item. Record the name and identity of the product, as well as the labeled or advertised price, for each item.

(c) Proceed through a check-out as if you were a customer and pay for the purchase. Obtain the original sales receipt, and compare the price charged with the labeled or advertised price for each item. Record the time of day, lane number, and the identity of the checker. Before leaving the store, determine if any errors have occurred. Identify yourself and inform the store management that a test purchase was conducted and report the results. (In many instances, the store will credit back all of the items and refund the test purchase money.) Record the information on the test report form and determine the cause of the error (e.g., operator error, mislabeling, or incorrect price sign).

7.4.2. Alternative Procedure - Consumer Complaints. – Complaints can be investigated by using any of the test procedures described above or by verifying only the price of the item or items subject to the complaint. If the complaint is valid, you can limit your inspection to the items described in the complaint or you may conduct a complete inspection.

7.4.3. Evaluation of Results. – The errors for items verified using these procedures should be evaluated according to Sections 9. Evaluation of Inspection Results and 10. Accuracy Requirements.

Section 8. Documentation of Findings

Several examples of Model Price Verification Reports are contained in pages 219 to 224. These forms were developed so that you only have to record the items found with price errors.

(a) Record errors and provide information on the cause, if determined. Indicate if the errors are considered to be violations, if stop-sale orders were issued, or if the violation was corrected.

(b) Notices of violations or other significant comments (e.g., warnings or violations ordered corrected) should always be included on the test form.

(c) Cash register receipts on verified items should be retained and attached to the inspection report as evidence.

(d) Printed advertisements and sales flyers should be retained and attached to the inspection report when errors are found in these categories.

Section 9. Evaluation of Inspection Results

9.1. Definition of Errors. – An error found to result from any of the following causes should not be considered a violation for enforcement purposes:

(a) An intentional undercharge if documentation or confirmation of the date and time of the price change is provided at the time of the inspection.

(b) An error caused by a mistake made in any kind of advertisement (e.g., newspaper, printed brochure, or radio or television advertisement) if the store has placed a notice adjacent to the item indicating that a mistake occurred in the advertisement.

(c) An error obviously caused by a price label that is missing or that has fallen off the shelf, or the item or the price label or sign has obviously been relocated by an unauthorized person.

(d) A "not-on-file" item is not an error unless you determine that the price "charged" for the item is incorrect (e.g., by conducting a test purchase or by asking the check-out clerk to determine the price of the item using the store's documented or customary procedures. If the price determined is incorrect, it is considered an error.)

NOTE: *It is recommended that you work with the store representative to identify the cause of any error and note the problem/cause on the report. This may not change your findings, but will help to identify problems related to* *staff errors, failure to follow through on established store pricing procedures, data entry errors, or failure of management to provide correct written data, etc. The supporting information will help with enforcement decisions as well as in-house monitoring of product pricing.*

9.2. Computing Sample Errors. – The following formulas are used to determine sample error and the overcharge to undercharge ratio:

(a) Adjust the total sample by subtracting any items or errors specified in 9.1. Definition of Errors.

(b) To compute the sample error, divide the number of errors by the total sample size to obtain the error in percent.

For example: a sample of 100 items is verified; 3 overcharges and 1 undercharge are found for a total of 4 errors:

4 ÷ 100 = 4 % sample error.

(c) To compute the ratio of overcharges to undercharges (used on large samples and in follow-up activities), total the overcharges/undercharges and compare the numbers:

3 overcharges/1 undercharge = a 3 to 1 ratio.

Section 10. Accuracy Requirements

10.1. Accuracy Requirements. – Accuracy information, based on a percentage of errors found in a sample and the ratio of overcharges to undercharges, constitutes useful criteria for evaluating the "pricing integrity" of the store. Both overcharges and undercharges should be considered as errors in taking lower level enforcement actions since (1) either type of error misrepresents the price of the item; and (2) the occurrence of any error in a randomized sample may indicate poor pricing practices that would result in errors where additional items were sampled. For higher levels of enforcement only overcharges are considered.

10.2. Accuracy. – The accuracy requirement for a sample must be 98 % or higher to "pass" a single inspection. See Column 4, Accuracy Requirements, in Table 1. Samples, Sample Collection, and Accuracy Requirements.

10.3. Ratio of Overcharges to Undercharges. – With large sample sizes, overcharges should not exceed the undercharges. A high rate of overcharges to undercharges (2 to 1, or 3 to 1) may indicate systematic problems with a store's pricing practices.

NOTE: *As the history of store compliance develops, the number of overcharges and undercharges may be evaluated to determine if systematic errors or other problems exist. This ratio should be maintained when at least 10 errors are found over several inspections, or in a single large sample size (e.g., the results of several 100-item inspections collected over a period of time or if 1000 items are sampled in one inspection.)*

Table 2. Price Errors
(This table shows the percentage of errors in different sample sizes)

Percentage of Errors
Sample Size

No. of Errors	25	50	100	150	200	300
1	4 %	2 %	1 %	0.67 %	0.50 %	0.33 %
2	8 %	4 %	2 %	1.33 %	1.00 %	0.67 %
3	12 %	6 %	3 %	2.00 %	1.50 %	1.00 %
4	16 %	8 %	4 %	2.67 %	2.00 %	1.33 %
5	20 %	10 %	5 %	3.33 %	2.50 %	1.67 %
6	24 %	12 %	6 %	4.00 %	3.00 %	2.00 %
7	28 %	14 %	7 %	4.67 %	3.50 %	2.33 %
8	32 %	16 %	8 %	5.33 %	4.00 %	2.67 %
9	36 %	18 %	9 %	6.00 %	4.50 %	3.00 %
10	40 %	20 %	10 %	6.67 %	5.00 %	3.33 %

NOTE: *Random pricing errors are to be expected, but the ratio of overcharges to undercharges will rarely be exactly 1 to 1 (e.g., of 10 errors, 5 overcharges and 5 undercharges); the ratio will likely vary both ways over several inspections. If a store has more overcharges than undercharges (e.g., 2 to 1, or 3 to 1), it may indicate that the store is not following good pricing practices, but enough errors must be present in order to make this determination. (Consider the example of 12 pricing errors consisting of 8 overcharges and 4 undercharges: the ratio of overcharges to undercharges is 2 to 1. Similarly, 10 pricing errors consisting of 6 overcharges and 4 undercharges correspond to a ratio of 1.5 to 1; since all decimal values are truncated to whole numbers, 1.5 is truncated to 1, and the ratio becomes 1 to 1.)*

The one-to-one ratio should be applied to any sample size if at least 10 errors are present. For example, if 1000 items are verified and 10 items are found in error, the sample has an accuracy of 99 %. However, if 9 of the 10 errors are overcharges (i.e., a ratio of 9 overcharges to 1 undercharge), the store should be considered to have poor pricing practices or other problems; if 100 items are verified and a 90 % accuracy is found, 10 items in error not meeting the overcharge to undercharge ratio can be used in enforcement action as evidence of poor pricing practices.

Section 11. Enforcement Procedures

11.1. Enforcement Steps.

(a) Compliance is based on the accuracy found on a sample collected according to this procedure.

(b) Errors should be corrected immediately, or if the correction cannot be made immediately, a stop-sale order shall be issued before you leave the business. If the errors are not corrected in your presence, a follow-up inspection may be made later in the day or the following day to ensure the store has corrected the error. If a store fails to correct the error by that time, higher level enforcement action should be taken.

(c) Enforcement action for large monetary errors on individual items, confirmed overcharges on items verified in response to complaints, or errors found on follow-up inspection of items ordered corrected, should be taken independently from any sample, giving consideration to the magnitude of the violation, corrective action by the establishment, and any other relevant information. Action may be initiated at any time in the inspection process based on the facts of the individual case.

(d) Overcharges and undercharges are used to determine lower levels of enforcement actions, but higher levels of enforcement action (e.g., fines or penalties) are taken only on the overcharges found in the sample.

(Amended 2001)

NOTE: *Many computer systems do not allow for the immediate correction of errors in the database. Downloading information throughout the day may not be possible. Therefore, for the purposes of this section, "immediate" correction of errors may entail the removal or correction of problem signs, manually changing marked prices, or communicating notice of the corrected price to all applicable stores through facsimile, e-mail, or any other appropriate medium to ensure that consumers are charged the correct price.*

11.2. Model Enforcement Levels.

These recommendations do not modify the enforcement policy of any jurisdiction unless adopted by that jurisdiction.

(a) **Ninety-Eight Percent or Higher.** – If price accuracy is 98 % or higher on a sample of 50 or more items, and if overcharges do not exceed undercharges on sample sizes of 100 or more items, and the store is on a normal inspection frequency:

(1) a notice of noncompliance is issued on violations, and the store is maintained on a normal inspection frequency; or

(2) if the store is on increased inspection frequency, it remains on this frequency until inspection results conform to Terms of Increased Inspection Frequency.

(b) **Less Than Ninety-Eight Percent.** – If price accuracy is less than 98 % on a sample of 50 or more items and if overcharges do not exceed undercharges on large sample sizes, and the store is on normal inspection frequency:

(1) A notice of noncompliance is issued and the store is placed on an increased inspection frequency.

(2) A second inspection should be conducted within 30 business days. If the price accuracy then is not 98 % or higher, a warning is issued.

(3) A third inspection should be made within 60 business days. If the price accuracy is again less than 98 %, higher level enforcement action should be taken.

(Amended 2001)

If the store is on increased inspection frequency, a warning should be issued and the store re-inspected within 30 business days. If price accuracy is less than 98 %, higher levels of enforcement action should be taken.

(Amended 2001)

Examples:

For the 100-item sample size:

- If 100 items are verified and three overcharges are found in the sample, the error rate is 3 %. In this example, higher levels of enforcement action should be taken.
- If 100 items are verified and three overcharges and two undercharges are found, the error rate on the sample is 5 %, but overcharges are 3 %. In this example, higher levels of enforcement action should be taken.
- If 100 items are verified and two overcharges and three undercharges are found, the error rate is still 5 %, but overcharges are only 2 % of the sample. In this example, a lower level enforcement action would be taken.

(c) **Terms of Increased Inspection Frequency.** – When a store is on increased inspection frequency, it shall remain at that frequency until two consecutive inspections reveal an accuracy of 98 % or higher.

(d) **Higher Levels of Enforcement Action.** – Overcharges and undercharges are used to determine lower levels of enforcement actions, but higher levels of enforcement action (e.g., fines or penalties) are taken only on overcharges. A store's history of error rates, the time it takes a store to correct the errors, the difference in inaccuracy rates found between "regular" and "sale" priced items, the ratio of overcharges to undercharges, a record of valid consumer complaints, and the magnitude of the error(s) may be used to support enforcement action.

Section 12. Post-Inspection Tasks

(a) You should meet with the store representative to review your findings. Have the inspection report completed prior to the meeting and be prepared to briefly summarize your findings and recommended actions, and provide a copy to the store representative.

(b) Return borrowed safety, sanitation, and/or test equipment.

(c) If you removed items from display, ensure that the items are returned to their proper location on the store shelves unless the representative requests to have the items returned by a store employee, which is permitted.

(d) Advise the representative of your findings. Explain any violations and errors. Explain any orders issued and be sure the individual acknowledges understanding of what corrective action is expected, if any.

(e) If necessary, describe the implications of the inspection results and advise the store of the action that you intend to take. If an increased inspection frequency is called for due to the accuracy level found during the inspection, advise the firm that re-inspections will be made, but do not indicate when they may occur.

Section 13. Supervisory Activities

13.1. Baseline Surveys. – Price verification programs require management support so that the program's objectives and desired benefits can be incorporated into the enforcement agency's work plans and budget. Surveys to measure pre- and post-implementation accuracy should be used to establish a base from which to measure whether a cost/benefit has been obtained.

13.2. Follow-up Inspections. – Inspections that reveal errors exceeding the accuracy requirements recommended above must include follow-up action to ensure that the store fulfills its obligations regarding accurate prices.

13.3. Management Information Systems. – To ensure adequate control and follow-up, a database should be established in each jurisdiction to provide information on every store, including:

For stores:

- store name
- address
- telephone
- type of store
- frequency of inspection
- sample size
- accuracy
- number of overcharges
- dollar value of overcharges
- number of undercharges
- dollar value of undercharges
- average money value of undercharges
- ratio of overcharges to undercharges

For program review:

- total number of undercharges
- total dollar value of undercharges
- average dollar value of undercharges
- percent undercharges of total
- ratio of overcharges to undercharges
- total error in percent
- accuracy levels of stores
- store type
- total stores tested
- total stores tested (each type)
- total items tested
- total number of overcharges
- total dollar value of overcharges
- average dollar value of overcharges
- percent of overcharges of total

Section 14. Model Forms for Price Verification Inspections

These models can be used to develop formal report forms, or they can be copied and used as worksheets for conducting inspections:

(a) **Sample Tally Sheets:** These forms can help you keep track of the number of items verified. They provide spaces to record the item's display location (e.g., aisle or department), a description of the item, and the shelf or advertised price. The worksheets are set up for the stratified sample collection described above to help identify the types of products to select.

(1) Price Verification Tally Sheet – Food Stores. (See page 219.)

(2) Price Verification Tally Sheet – Department Stores. (See page 220.)

(b) **Model Inspection Form I:** This can be used to document violations and record findings. A completed sample is provided.

(1) Price Verification Report I – sample blank form. (See page 221.)

(2) Price Verification Report I – completed sample form. (See page 222.)

(c) **Model Inspection Form II:** This can be used in stores where a hand-held scanning device is not available, or when it is inconvenient to take items (e.g., a large ladder in a hardware store) to a check-out register to verify the price. You can record an identity, the UPC or PLU code, and advertised price so that you can manually enter the codes to verify the price. The form can also be used to record findings. A completed sample is provided.

(1) Price Verification Report II – sample blank form. (See page 223.)

(2) Price Verification Report II – completed sample form. (See page 224.)

Price Verification Tally Sheet - Food Stores

	Item	Location	UPC/Identity	Shelf Price
"End of Aisle" or "Tie-In Display"	1.	1.	1.	1.
	2.	2.	2.	2.
	3.	3.	3.	3.
	4.	4.	4.	4.
"PLU or Coded" Items	6.	6.	6.	6.
	7.	7.	7.	7.
	8.	8.	8.	8.
	9.	9.	9.	9.
	10.	10.	10.	10.
"Advertised Sale" Items	11.	11.	11.	11.
	12.	12.	12.	12.
	13.	13.	13.	13.
	14.	14.	14.	14.
	15.	15.	15.	15.
	16.	16.	16.	16.
	17.	17.	17.	17.
	18.	18.	18.	18.
	19.	19.	19.	19.
	20.	20.	20.	20.
Items on "Special"	21.	21.	21.	21.
	22.	22.	22.	22.
	23.	23.	23.	23.
	24.	24.	24.	24.
	25.	25.	25.	25.
"Direct Store Delivery" Items	26.	26.	26.	26.
	27.	27.	27.	27.
	28.	28.	28.	28.
	29.	29.	29.	29.
	30.	30.	30.	30.
"Randomly Selected" Items	31.	31.	31.	31.
	32.	32.	32.	32.
	33.	33.	33.	33.
	34.	34.	34.	34.
	35.	35.	35.	35.
	36.	36.	36.	36.
	37.	37.	37.	37.
	38.	38.	38.	38.
	39.	39.	39.	39.
	40.	40.	40.	40.
	41.	41.	41.	41.
	42.	42.	42.	42.
	43.	43.	43.	43.
	44.	44.	44.	44.
	45.	45.	45.	45.
	46.	46.	46.	46.
	47.	47.	47.	47.
	48.	48.	48.	48.
	49.	49.	49.	49.
	50.	50.	50.	50.

Price Verification Tally Sheet - Department Stores

	Identity	Location	Advertised
"End of Aisle" or "Tie-In Display"	1.	1.	1.
	2.	2.	2.
	3.	3.	3.
	4.	4.	4.
	5.	5.	5.
"Advertised Sale" Items	6.	6.	6.
	7.	7.	7.
	8.	8.	8.
	9.	9.	9.
	10.	10.	10.
	11.	11.	11.
	12.	12.	12.
	13.	13.	13.
	14.	14.	14.
	15.	15.	15.
Items on "Special"	16.	16.	16.
	17.	17.	17.
	18.	18.	18.
	19.	19.	19.
	20.	20.	20.
	21.	21.	21.
	22.	22.	22.
	23.	23.	23.
	24.	24.	24.
	25.	25.	25.
"Randomly Selected" Items	26.	26.	26.
	27.	27.	27.
	28.	28.	28.
	29.	29.	29.
	30.	30.	30.
	31.	31.	31.
	32.	32.	32.
	33.	33.	33.
	34.	34.	34.
	35.	35.	35.
	36.	36.	36.
	37.	37.	37.
	38.	38.	38.
	39.	39.	39.
	40.	40.	40.
	41.	41.	41.
	42.	42.	42.
	43.	43.	43.
	44.	44.	44.
	45.	45.	45.
	46.	46.	46.
	47.	47.	47.
	48.	48.	48.
	49.	49.	49.
	50.	50.	50.

Price Verification Report I

Page ___ of ___

Inspection: [ ] 1st [ ] 2nd [ ] 3rd **Frequency:** [ ] Normal [ ] Increased **Type:** [ ] Stratified [ ] Automated [ ] Randomized **Complaint:** [ ]

Location of Test (Store Name, Address, County, ZIP Code)	Date:	Telephone:
	Manager:	Type of Store:

Identity, Brand Name, Item or Style Number	Number of Items, Size, Location in Store, or UPC Code	Offered Price	Price Charged	Error (±)
1.				
[ ] Stop Sale Issued [ ] Corrected	Comments:			
2.				
[ ] Stop Sale Issued [ ] Corrected	Comments:			
3.				
[ ] Stop Sale Issued [ ] Corrected	Comments:			
4.				
[ ] Stop Sale Issued [ ] Corrected	Comments:			
5.				
[ ] Stop Sale Issued [ ] Corrected	Comments:			
6.				
[ ] Stop Sale Issued [ ] Corrected	Comments:			
7.				
[ ] Stop Sale Issued [ ] Corrected	Comments:			
8.				
[ ] Stop Sale Issued [ ] Corrected	Comments:			

Inspection Results:

_____ (Sample Count) - _____ (#Not on File) = _____ (Adjusted Sample Count [ASC])

_____ (#Errors) ÷ _____ (ASC) = _____ (Error Percentage)

(Accuracy Percentage) = _____ % Overcharges/Undercharges Ratio = _____:_____

Inspector Name: ______________________________
Time In: ______________ Time Out: ______________
Comments/Remarks: ____________________________
____
____

Report Acknowledgement:
Name/Title: __________________________________
Comments/Remarks: ____________________________
____
____

Price Verification Report I (completed sample)

Page _1_ of _1_

Inspection: [√]1st [ ] 2nd [ ] 3rd **Frequency:** [√] Normal [ ] Increased **Type:** [√] Stratified [ ] Automated [ ] Randomized **Complaint:** [ ]

Location of Test (Store Name, Address, County, ZIP Code)	Date:	Telephone:
Barkers Food Store ***1361 Macon Street*** ***Belle, New Jersey 31756***	*3/10/95*	*(301) 555-4868*
	Manager: ***C. Barker***	Type of Store: ***Food Store***

Identity, Brand Name, Item or Style Number	Number of Items, Size, Location in Store, or UPC Code	Offered Price	Price Charged	Error (±)
1. ***Smith Cake Mix***	**32 oz. 313461346177**	**3.19**	**4.19**	**+1.00**
[ ] Stop Sale Issued [√] Corrected	Comments: ***Sale sign not removed***			
2. ***Natural Fruit Juice***	**1 Liter 617369345619**	**2.25**	**2.75**	**+.50**
[ ] Stop Sale Issued [√] Corrected	Comments:			
3. ***Clocks Soap***	**8oz. 936125376558**	**1.19**	**1.00**	**-.19**
[ ] Stop Sale Issued [√] Corrected	Comments:			
4.				
[ ] Stop Sale Issued [ ] Corrected	Comments:			
5.				
[ ] Stop Sale Issued [ ] Corrected	Comments:			
6.				
[ ] Stop Sale Issued [ ] Corrected	Comments:			
7.				
[ ] Stop Sale Issued [ ] Corrected	Comments:			
8.				
[ ] Stop Sale Issued [ ] Corrected	Comments:			

Inspection Results:

_100_ (Sample Count) - _0_ (#Not on File) = _100_ (Adjusted Sample Count [ASC])

_3_ (#Errors) ÷ _100_ (ASC) = _3_ (Error Percentage)

(Accuracy Percentage) = _97_ % Overcharges/Undercharges Ratio = _2_ : _1_

Inspector Name: _T. Price_

Time In: _8:15_ Time Out: _9:30_

Comments/Remarks: ____

Report Acknowledgement:

Name/Title: _Chris Barker_

Comments/Remarks: ____

Price Verification Report II

Page ___ of ___

Inspection: [ ] 1st [ ] 2nd [ ] 3rd **Frequency:** [ ] Normal [ ] Increased **Complaint:** [ ]

Location of Test (Name, Address, County, ZIP Code)	Date:	Telephone:
	Manager:	Type of Store:

Item/Size or Style Number	Offered Price	Price Charged	Price Error (±)	Item/Brand/Description/Code/Size	Offered Price	Price Charged	Price Error (±)
1. Identity: UPC/PLU: Comments:				11. Identity: UPC/PLU: Comments:			
2. Identity: UPC/PLU: Comments:				12. Identity: UPC/PLU: Comments:			
3. Identity: UPC/PLU: Comments:				13. Identity: UPC/PLU: Comments:			
4. Identity: UPC/PLU: Comments:				14. Identity: UPC/PLU: Comments:			
5. Identity: UPC/PLU: Comments:				15. Identity: UPC/PLU: Comments:			
6. Identity: UPC/PLU: Comments:				16. Identity: UPC/PLU: Comments:			
7. Identity: UPC/PLU: Comments:				17. Identity: UPC/PLU: Comments:			
8. Identity: UPC/PLU: Comments:				18. Identity: UPC/PLU: Comments:			
9. Identity: UPC/PLU: Comments:				19. Identity: UPC/PLU: Comments:			
10. Identity: UPC/PLU: Comments:				20. Identity: UPC/PLU: Comments:			

Inspection Results:

_____ (Sample Count) - _____ (#Not on File) = _____ (Adjusted Sample Count [ASC]) Stop-Sale Order Issued? [ ]

_____ (#Errors) ÷ _____ (ASC) =_____ (Error Percentage)

(Accuracy Percentage) = _____ % Overcharges/Undercharges Ratio = _____:_____

Inspector Name: ____________________

Time In: ____________ Time Out: ____________

Comments/Remarks: ____________________

Report Acknowledgement:

Name/Title: ____________________

Comments/Remarks: ____________________

Price Verification Report II (completed sample)

Page _1_ of _1_

Inspection: [√] 1st [ ] 2nd [ ] 3rd **Frequency:** [√] Normal [ ] Increased **Complaint:** [ ]

Location of Test (Name, Address, County, ZIP Code)	Date:	Telephone:
Mark Downtown Department Store ***11650 Main St.*** ***Alice, MN 61619***	***3/16/95***	***(614) 555-6146***
	Manager: ***Jim Chester***	Type of Store: ***Department Store***

Item/Size or Style Number	Offered Price	Price Charged	Price Error (±)	Item/Brand/Description/Code/Size	Offered Price	Price Charged	Price Error (±)
1. Identity: ***Sony Color TV*** UPC/PLU: ***38569*** Comments: ***Model 6136X***	***189.00***	***199.00***	***+10.00***	11. Identity: UPC/PLU: Comments:			
2. Identity: ***Moore Lawn Mower*** UPC/PLU: ***31619*** Comments: ***Shp with bagger***	***96.00***	***91.00***	***-5.00***	12. Identity: UPC/PLU: Comments:			
3. Identity: ***Taft Rake*** UPC/PLU: ***39916*** Comments: ***Not on file***	***8.99***			13. Identity: UPC/PLU: Comments:			
4. Identity: ***Calendar*** UPC/PLU: ***615191*** Comments:	***5.50***	***7.10***	***+1.60***	14. Identity: UPC/PLU: Comments:			
5. Identity: UPC/PLU: Comments:				15. Identity: UPC/PLU: Comments:			
6. Identity: UPC/PLU: Comments:				16. Identity: UPC/PLU: Comments:			
7. Identity: UPC/PLU: Comments:				17. Identity: UPC/PLU: Comments:			
8. Identity: UPC/PLU: Comments:				18. Identity: UPC/PLU: Comments:			
9. Identity: UPC/PLU: Comments:				19. Identity: UPC/PLU: Comments:			
10. Identity: UPC/PLU: Comments:				20. Identity: UPC/PLU: Comments:			

Inspection Results:

_50_ (Sample Count) - _1_ (#Not on File) = _49_ (Adjusted Sample Count [ASC]) Stop-Sale Order Issued? [ ]

_3_ (#Errors) ÷ _49_ (ASC) = _6_ (Error Percentage)

(Accuracy Percentage) = _94_ % Overcharges/Undercharges Ratio = _2_ : _1_

Inspector Name: _Tim Marlowe_
Time In: 10:25 AM Time Out: 4:45 PM
Comments/Remarks: ______

Report Acknowledgement:
Name/Title: _Jim Chester, Manager_
Comments/Remarks: ______

VI. NCWM Policy, Interpretations, and Guidelines, Section 2 Excerpts from NCWM Publication 3

Table of Contents

Section **Page**

NCWM Policy, Interpretations, and Guidelines

Introduction

This section of the handbook includes NCWM interpretations, policies, recommendations, inspection outlines, and information on issues that have come before the Conference. Several sections include information on federal requirements related to the uniform laws and regulations presented in the handbook. The purpose of this section is to assist users in understanding and applying the uniform regulations and to guide administrators in implementing new programs or procedures. The guidelines or recommendations provided should not be construed to redefine any state or local law or limit any jurisdiction from enforcing any law, regulation, or procedure (unless the section describes a specific federal regulation that preempts local requirements).

(Added 1997)

2.1.1. Weight(s) and/or Measure(s).

(L&R, 1985, p. 77)

The measuring elements of a point-of-sale system are "weights and/or measures." Errors in pricing when found in point-of-sale systems come under "Misrepresentation of Pricing" in the weights and measures law and are under the jurisdiction of weights and measures.

Background

A recommendation was made to change the definition of "weights and measures" in the Uniform Weights and Measures Law to specifically define a scanner or point-of-sale system as under weights and measures jurisdiction.

Several state representatives said that they had enforcement problems when a scanner or point-of-sale system was being used and when the price marked on an item (or on the shelf) was not the same as the price printed on the receipt. These officials believe that unless the law specifically defines these devices as "weights and measures," they have no jurisdiction over the devices' function.

The Committee disagreed. The NCWM Uniform Weights and Measures Law has a section that forbids the practice of a different price on the retail shelf as compared with the price provided by a scanner. Section 15 of the Uniform Weights and Measures Law reads:

> *No person shall misrepresent the price of any commodity or service sold, offered, exposed, or advertised for sale by weight, measure, or count, nor represent the price in any manner calculated or tending to mislead or in any way deceive a person.*

This section (plus Section 14 forbidding misrepresentation of quantity), if enacted by a state, already provides enforcement authority over scanners and point-of-sale systems.

In addition, the Committee does not want to set a precedent by listing by name the types of devices that might be considered weights and measures devices. This might provide a potential "loop-hole" for those devices not specifically listed. Finally, the Committee members pointed out that it is the human element (the person reading in data or receiving price updates) that introduces the discrepancies in shelf and receipt prices rather than any inherent incapability of the reading device or scanner. Therefore, it is much more effective to forbid the practice of mispricing rather than focus on a single device or apparatus as the means for obtaining compliance.

2.1.2. Section 19(a), Identity.
(L&R Committee, 1986, p. 143)

Packaged food not containing meat or poultry does not have to have an identity statement if the identity of the commodity can easily be identified through the wrapper or container.

Background

Virginia Weights and Measures recommended revision to Section 19(a) of the Uniform Weights and Measures Law (UWML) to eliminate the exemption of an identity statement from packages when the item "can easily be identified through the wrapper or container." The Committee is of the opinion that there is merit in retaining the language in Section 19(a) of the Uniform Law. Packages of fresh product packaged in a retail establishment are considered to be packages as long as a price is attached. If the exemption were eliminated, such packages instead of being marked, for example, "12/89 cents" would have to be marked "lemons, 12/89 cents." It was argued that there could be a problem in deciding whether or not a commodity could "easily be identified" (such as might occur in an ethnic specialty grocery or with an exotic produce item). In researching the issue, the Committee has determined that Title 21, Section 101.100(b)(3) of the Code of Federal Regulations specifically exempts the food identity statement from having to appear ". . . if the common or usual name of the food is clearly revealed by its appearance." Since no specific problems of enforcement were brought to the attention of the Committee concerning this issue, the Committee recommends no change to Section 19(a) at this time. However, the Committee recommends that Section 3.1. and 4. of the Uniform Packaging and Labeling Regulation be noted as follows:

> *Section 19(a) of the Uniform Weights and Measures Law, and 21 CFR 101.100(b) (3) for non-meat and non-poultry foods, specifically exempt packages from identity statements if the identity of the commodity "can easily be identified through the wrapper or container."*

2.1.3. Definition of Net Weight.
(L&R, 1987, p. 123)

1. It is the intent of this definition to include truck-loads of commodities, not just packages ("containers").

2. It is not the intent to define the net weight of packaged goods as requiring dry tare (". . . excluding . . . substance(s) not considered to be part of the commodity" could just as well be interpreted as excluding liquids not considered part of the commodity at the time of sale).

3. It is also the intent to permit more specific definitions as the occasion warrants (". . . material(s) . . . not considered . . . part of the commodity" might include dirt or "foreign material" in a commodity).

2.1.4. Offenses and Penalties, Sale of an Incorrect Device.
(L&R, 1987, p. 124)

A jurisdiction seeking to enforce the provision of the Uniform Weights and Measures Law that prohibits the sale of an incorrect device would have to show that the seller knowingly sold or offered for sale for use in commerce an incorrect weight or measure. Under Section 22, a seller would not be responsible for actions taken by the purchaser or distributor, in which the seller did not participate or have prior knowledge. Thus, the seller would not be liable:

(1) if a purchaser or distributor modified a scale obtained from a seller; or

(2) if a scale were used in trade after the seller informed the purchaser that the scale was not appropriate for that use.

In cases such as those noted above, the Committee feels that the seller would be protected from prosecution. Only sellers who knowingly violate the provision would be subject to prosecution.

2.1.5. Weight: Primary Mill Paper.
(L&R, 1990, p. 81)

Interpretation

Non-consumer sales of "primary mill paper" were discovered by weights and measures officials to be labeled and invoiced on what was called a "gross weight" basis. Primary mill paper is produced for commercial or industrial companies for subsequent additional processing, such as paper for newspaper or magazine publishers or sanitary tissue manufacturers. The primary mill paper is cut from "parent rolls" but is still a commercial-sized item weighing from several hundred to several thousands of pounds.

The key to understanding the longstanding trade practice is that the purchaser of such paper specifies not only the quality of the paper being purchased, such as the thickness, surface coating, etc., but the purchaser also specifies the core around which the paper is to be wound, the type of overwrap, the number of overwraps, and such other requirements that will ensure receipt of the primary mill paper in proper condition for subsequent processing. The weight of the core and wrapping is approximately 1 % of the gross weight. It is recycled by the purchasers in their own or other paper recovery or reuse systems.

Having reviewed the practices in the industry in the specification and purchasing of primary mill paper, the Committee concludes that the true product is the paper plus the packaging (in order to assure maintenance of quality) and an appropriate core (to ensure a fit on the recipient's equipment). Therefore, in the Committee's opinion, the sale of primary mill paper is not at all on a gross weight basis. This is and has been a misnomer. The true identity of the purchased product has been misunderstood by weights and measures authorities, further compounded by the industry use of the term "gross weight." The product is the primary mill paper plus the core and overwrap specified by the purchaser.

The Committee, therefore, believes that the industry should review its invoicing and labeling to clarify that the weight of the specified product is the weight of the primary mill paper, core, and overwrap. Although this weight is the gross weight of the entire item as produced and shipped, it is the net weight of the item as specified by the purchaser.

This interpretation applies only to primary mill paper and is not intended to be applied to all non-consumer products ordered by specification; it is a narrow interpretation applying to the specific method of sale in this trade where the service of packaging and the packaging is part of the purchase.

2.2.1. Gift Packages.
(Resol. 1975, p. 237)

See also Interpretation 2.2.8.

Interpretation

Seasonal gift packages are often put up in retail stores in baskets and other decorative containers using cellophane or other clear flexible wrap to enclose a number of similar or dissimilar prepackaged items (for example: cheese, jellies, sausages, wine, fruit, etc.). The resulting combination or variety package must have a legally conforming label including the net contents statement.

2.2.2. Sand.
(L&R, 1978, p. 151)

Interpretation

Sand put up in permanent wooden bins is a consumer package and must be labeled with all mandatory information as required by the Uniform Packaging and Labeling Regulation.

Background

The State of Hawaii raised the issue of the sale of sand in permanent wooden bins and sold by price per cubic measure. The Committee agrees with Hawaii that the sale of sand in this manner is subject to the Uniform Packaging and Labeling Regulation, under the definition of "Consumer Package" (Section 2.2. of the Uniform Packaging and Labeling Regulation) and that no further action is needed.

2.2.3. Sold by $^{4}/_{5}$ Bushel.

(L&R, 1974, p. 220)

Interpretation

The trade practice of crating citrus fruit in $^{4}/_{5}$ bushel units is a long-standing one. It is not intended to be a consumer package. If offered as a consumer package, the general consumer usage and trade custom in the particular state would have to be explored:

Section 6.10.(b)(1) of the Uniform Packaging and Labeling Regulation would permit a declaration employing different fractions in the net quantity declaration other than those permitted under Section 6.10.(b) if there exists a firmly established practice of using $^{4}/_{5}$ bushel in consumer sales and trade custom.

Background

It has been called to the attention of the Committee that certain commodities are being sold to consumers in "unacceptable" fractional units of dry measure in violation of Section 6.10. of the Uniform Packaging and Labeling Regulation. Specifically, the Committee has been asked for an interpretation as to whether the packaging of oranges in a $^{4}/_{5}$ bushel, which is later sold unweighed to a consumer, is a violation of the binary submultiple principle as implied in Section 6.10.(b). Some Committee members asserted that a clear exception exists under Section 6.10.(b)(1) which applies to this long established tradition of crating citrus fruit in $^{4}/_{5}$ of a bushel. Approximately 85 % of this fruit is sold by this trade practice. Additionally, it was asserted that the packager never intended the $^{4}/_{5}$ bushel to be a consumer package, but if the $^{4}/_{5}$ bushel of citrus fruit is sold to consumers, this would be a matter between the appropriate state or local official and the retailer.

The consensus of the Committee is that this action of the packagers is not in violation of the indicated section.

2.2.5. Lot, Shipment, or Delivery.

(L&R, 1981, p. 95)

Policy

The requirements for the average package net contents to meet or exceed the labeled declaration may be applied to production lots, shipments, or deliveries. Shipments or deliveries are smaller collections of packages than production lots that may or may not consist of mixed lot codes.

Emphasis in inspection activities should be placed on warehouse and in plant testing without neglecting retail consumer protection.

Background

The Committee heard a petition from the California Brewers Association to define a lot as:

> *...a selection of containers under one roof produced by a single company of the same size, type and style, manufactured or packed under similar conditions with a minimum number to be equivalent to one production line shift.*

The intention of the petition is to focus Weights and Measures enforcement on production lots as opposed to small collections of packages on retail shelves, because the production lot is under the control of the packager.

An alternative proposal was made that would require mingling of lot and date codes in package inspection at warehouse locations.

The Committee has reviewed the proposals in light of Section 7.6. and Section 12.1. of the Uniform Packaging and Labeling Regulation which refers to "shipment, delivery, or lot." If the petition is approved, the terms "shipment" and "delivery" would have to be dropped from this Uniform Regulation.

The Committee recognizes the inherent value of in-plant and warehouse inspection and is of the opinion that, wherever possible, such inspections should be carried out. At the same time, the Committee recognizes the need for the state and local weights and measures officials to protect the consumer at the level where the ultimate sale is made. Therefore, the Committee recommends no change to the Uniform Regulation.

The Committee looks forward to the work of the Special Study Group on Enforcement Uniformity of the NCWM which will be exploring the mechanisms that might be instituted to make in-plant inspection workable.

2.2.6. Aerosols and Similar Pressurized Containers.

(L&R, 1976, p. 248)

See also Guideline 2.2.7.

Interpretation

It is the opinion of the NCWM that an FDA opinion as expressed in the Fair Packaging and Labeling Act Manual Guide FDA 7563.7, not objecting to volume declarations on aerosol products, does not supersede or preempt state requirements that aerosols be labeled by net weight.

Background

The Department of Commerce through the Office of Weights and Measures of the National Institute of Standards and Technology, under its statutory responsibility for "cooperation with the states in securing uniformity in weights and measures laws and methods of inspection," developed Section 10.3.

> **10.3. Aerosols and Similar Pressurized Containers.** – The declaration of quantity on an aerosol package and on a similar pressurized package shall disclose the net quantity of the commodity (including propellant), in terms of weight, that will be expelled when the instructions for use as shown on the container are followed.

Several states, which are among the 32 that have adopted the Uniform Packaging and Labeling Regulation, indicated that pressurized cans were currently being marked by volume rather than by weight as required above. Industry representatives indicated that according to the FDA, they are permitted to mark this type of container by volume and that for competitive purposes they will continue to do so. The NCWM was asked to contact FDA and inform them that a declaration of volume on pressurized containers is not acceptable to the states since it cannot be verified.

A meeting was requested to express NIST/NCWM's concern over the FDA position on quantity of contents declarations on aerosols, which is found in the Fair Packaging and Labeling Act (FPLA) Manual Guide FDA 7563.7. This Guide states that in the past, the FDA has not objected to the use of units of volume to declare the net contents of aerosol preparations that would be liquid if not combined with the propellant and a net weight statement in avoirdupois units for products that would be solids if not combined with a propellant. The FDA was asked to modify its position to provide that existing state regulations (concerning aerosol quantity of contents declarations) are not superseded by FDA Guidelines. FDA officials stated that the FDA would consider the request, but it did not appear at the time of the Interim Meetings that the FDA would make any statement to modify its position without following its administrative procedures and permitting interested parties to exhaust every element of due process.

One industry representative stated that there has been a good deal of concern that fluorocarbon propellants may in the long run cause the partial destruction of the ozone layer in the upper atmosphere surrounding the earth, and that the diminution of the ozone layer would have adverse effects on human health. Therefore, they have converted to new formulations which eliminate fluorocarbon propellants. As a result of this conversion to a non-fluorocarbon propellant system, which uses a propellant with a much lower density than that of the usual fluorocarbon propellants, continued use of a weight measure would be highly misleading to the consumer. Therefore, some spray labels have been changed so as to denote the contents in terms of fluid measure, rather than in terms of weight measure.

The industry representative stated that if manufacturers were to be required to use weight measure, consumers would be deceived into buying products, such as hair spray, with large amounts of fluorocarbon that vaporizes before it reaches the hair. Consumers prefer products with a large amount of base. Industry further indicated that they wanted to avoid a confrontation with the states over this issue and believe that the matter can readily be resolved without the need for litigation. Although the use of fluid measure on the principal panel will give consumers the most helpful information at the point of purchase, the industry would have no objection to putting the net weight on the back of the label.

The Committee wants to commend FDA for their interest in this matter and the manufacturers who seek to improve their product and its labeling information. The Committee is also encouraged to work with all interested parties to resolve this issue. However, the Committee does not believe that mere guidelines can preempt a Uniform Regulation developed under the technical authority of the federal agency delegated by Congress and adopted by the states through its representatives, no matter how broad the preemptive clause of an act might be. Additionally, the Committee cannot support open and notorious violations of state regulations where those violations occurred prior to bringing the issue before the Conference. Therefore, the Committee believes that NCWM should support a firm stand by the states that their regulations must be respected.

2.2.7. Aerosol Packaged Products.

(Liaison, 1979, p. 239)

See also Guideline 2.2.6.

Policy

The NCWM recommends all aerosol packages be labeled by net weight. FDA permits volume declarations. The NCWM has requested the FDA to change its regulations and revise its interpretation of these regulations.

Substance of Petition

The NCWM petitions the FDA to make the necessary changes to their regulations and interpretation of 21 CFR 101.105(g) as appearing in the FDA Fair Packaging and Labeling Manual Guide, 7563.7 pertaining to the quantity of contents declaration on aerosol packaged products. It is requested that the net quantity statement on aerosol packaged products or similar pressurized packages be made in terms of net weight only. The reasons for recommending such changes are as follows:

1. Net quantity labeling of aerosol packaged products in terms of net weight is a firmly established trade practice for such products.

2. Net quantity labeling of aerosol packaged products in terms of volume is difficult (if not impossible) to verify with consumer verification methods or by conventional package inspection methods. State or local enforcement action is discouraged by such labeling.

3. Since the labeling of aerosol packaged products by volume cannot be compared with the labeling of such products in terms of net weight, labeling in terms of volume and weight inhibits value comparisons and causes consumer confusion with respect to the quantity of product the consumer is buying and can be a form of deceptive labeling.

4. Uniformity between all state and federal regulations is highly desirable for both enforcement and fair competition in the marketplace. The Uniform Packaging and Labeling Regulation and the FTC and EPA Regulations require net quantity labeling of aerosol packaged products in terms of net weight.

2.2.8. Variety and Combination Packages.
(L&R, 1982, p. 149)

See also Guideline 2.2.1.

Interpretation

(a) Seasonal gift packages are "variety packages" within the meaning of the Uniform Packaging and Labeling Regulation if they contain "reasonably similar commodities" (such as various fruits). They are "combination packages" if they contain "dissimilar commodities" (such as wine, fresh fruit, and jellies). Variety package labels must declare the total quantity in the package. Combination package labels must declare a quantity declaration for each portion of dissimilar commodities.

(b) The example provided with Section 10.6., Variety Packages, of the Uniform Packaging and Labeling Regulation, shows a total quantity declaration and individual declaration for each type of commodity. The individual declaration is not required but is encouraged.

Background

The Committee reviewed Section 10.5 and Section 10.6 of the Model Packaging and Labeling Regulation in order to determine the need for further clarification. Several questions have arisen over the years with respect to:

(1) What are the net contents labeling requirements for seasonal gift packages composed of varying types of commodities or goods all combined into one package?

(2) Is the example provided in Section 10.6. entirely in keeping with the declaration requirements? (This section requires that total net contents be declared, but the example shows both total and individual net contents.)

The Committee believes that there is no need to modify these sections, but the discussions below may serve as guidance to enforcement officials and packagers on these sections.

Concerning labeling requirements for seasonal gift packages, it must first be determined what the individual units comprising each package are. The following examples are possibilities:

(a) individual packages of sausage, individual packages of cheese;

(b) several kinds of fruit of different weights; and

(c) several kinds of fruit, bottle of wine, several packages of cheese.

Examples (a) and (c) above are combination packages and should be labeled with net quantities of each unit or type of unit. It is possible to combine fruit net weight (or count if appropriate) as one declaration, cheese net weight as a second declaration, etc.

Example (b) above is a variety package and must be labeled with the total net weight or count (as appropriate) of fruit in the package. It is also reasonable for packagers to include, for full consumer information, a declaration of the individual net contents of each type of package or item in the gift package although this latter declaration is not required (e.g., 1 lb bananas, 3 pears). This is also the key to the second question asked above concerning the example provided in Section 10.6.; that is, although a declaration of individual item net contents is not required, packagers are encouraged to provide additional information wherever useful to the consumer.

2.2.9. Textile Products.

(L&R, 1977, p. 215)

Interpretation

(a) When a range of widths (e.g., 58/60) appears on the label of bolts or rolls for yard goods, enforcement action should be taken whenever the action width falls below the lesser of the two widths given as the range (in the example above, when the fabric width is less than 58 in).

(b) Section 10.9.3. Textiles: Variations from Declared Dimensions of the Uniform Packaging and Labeling Regulation is not to be interpreted as providing tolerances. The average requirement must be met. The average quantity of contents of a lot, shipment, or delivery must equal or exceed the declared dimensions. Dimensions of individual packages of textiles may vary as much as Section 10.9.3. permits, but the average requirement must still be met.

Background

The State of California and the American Textile Manufacturers Institute asked the NCWM Laws and Regulations Committee and the National Institute of Standards and Technology to assist in the resolution of two textile-product issues. In the first issue California asks for help in correcting a short measure condition, apparently a nationwide problem, which has been found in the packaging and labeling of textile yard goods put up on bolts or rolls.

The problem is outlined as follows:

1. Approximate width measurements are being used by some manufacturers in their label declarations.

   Example:
   58/60 in (inch) width.

2. Label declarations are false and misleading in that actual amounts are less than the quantity represented on the label.

3. Section 10.9.3. of the Uniform Packaging and Labeling Regulation is extremely vague as to its intent and true meaning. Are the substantial variations (3 % and 6 %); (6 % and 12 %) permitted as product tolerances, or are they maximum unreasonable minus and plus errors to be allowed when sampling the product for quantity when using Handbook 67?

California favors the repeal or clarification of Section 10.9.3. and suggests amending Section 10.9.2.(k) to read:

> *The quantity statement for packages of textile yard goods packaged on the bolt or roll for either wholesale or retail shall state its net measure in terms of yards for the length and width of the item, or its net weight in terms of avoirdupois pounds or ounces, or in terms of their metric equivalent.*

During the Interim Meetings, a representative of the American Textile Manufacturers Institute (ATMI) informed committee members that the proposal to identify the width of yard goods with a single measurement (as opposed to a range) would be given serious consideration by their members, after which a recommendation will be finalized and submitted to the Laws and Regulations Committee.

After the Interim Meetings, the National Home Sewing Association said that if a single width declaration is required, the following could result:

(a) No change in manufacturing process would be effectuated; only the size declaration on bolts would be changed.

(b) Short measure problems could be created because consumers would look for the fabric to be exactly the stated width. Because the manufacturing processes were not changed, the width is actually the same as it was with the range declaration.

(c) Increased cost to manufacturers would result. One loom is used for many different fibers now; a single width declaration could create a need for many looms for each of the different fibers, thereby imposing "pass-along" costs to consumers.

(d) Consumer deception would be fostered in that a single declaration implies actual measurement.

California officials state that roll or bolt fabric should be labeled accurately with a single declaration. Additionally, they believe that industry does have enough shrinkage data on fibers used in the manufacturing processes, and thus could provide accurate measurement declaration on finished fabrics or materials.

The Committee believes that accurate quantity information should be provided on consumer products; however, no labeling changes should be required until patterns and yard goods are marketed in metric units. At that time, all measures shall be singularly stated (eliminating dual numbers) and, until that time, any products where size declaration is a range and found to be less than the smaller of the range declaration shall be subject to enforcement action. For example, a product marked "58 to 60 in" and found to be less than 58 inches should be considered to be in violation of weights and measures laws and/or regulations.

Additionally, the Committee affirms that the intent of the Variations from Declared Dimensions permitted in Section 10.9.3. in no way eliminates the requirement that quantity declarations for textiles must, on the average, not be less than declared declarations.

2.2.10. Yarn.
(L&R, 1983, p. 153)

Interpretation

The appropriate net contents declaration for yarn is weight.

Background

A consumer has requested that the net quantity statement for yarn be changed from weight to length. The proposal is based on the consumers' use of the product, darker colors often weigh more per unit of length. Therefore, they found that a lighter color yarn will "go farther" in craft applications than a darker yarn; consumers indicate that it is difficult to predict how much yarn of varying colors to purchase based on a weight declaration. The Committee is sympathetic to the request but must support existing labeling requirements for several reasons.

Yarn, by nature, is extremely stretchy; in order to label yarn by length, a specified tension would have to be applied in order to make any repeatable length measurement. Such a tension would have to be agreed upon by all the yarn manufacturers, and they would have to apply to compliance testing of product by weights and measures officials. Even if this tension "standard" were negotiated and decided upon, it would have little real meaning in use by needle crafters, knitters, and others. The tension applied to yarn in use varies from user to user and from application to application; therefore, the length also varies. Not only does dyeing yarn change the weight, dyeing also changes the length of yarn. For these reasons, industry representatives also support the requirements as they presently are written in the Uniform Packaging and Labeling Regulation.

The Committee recognizes the difficulty of working with this product and suggests that users of yarn consider buying an excess of the yarn over what is expected to be used in any application. The consumers should find out before purchase if, after finishing the product, they can return the unopened skeins to the retailers from whom the skeins were purchased.

2.2.11. Tint Base Paint.

(L&R, 1986, p. 146)

Section 11.23. of the Uniform Packaging and Labeling Regulation currently permits tint base paints (paints to which colorant must be added prior to sale) to be labeled in terms of the volume (a quart or gallon) that will be delivered to the purchaser after addition of the colorant only if three conditions are met:

1. "the system employed ensures that the purchaser always obtains a quart or a gallon";

2. "a statement indicating that the tint base paint is not to be sold without the addition of colorant is presented on the principal display panel;" and

3. "the contents of the container, before the addition of colorant, is stated in fluid ounces elsewhere on the label."

2.2.12. Reference Temperature for Refrigerated Products: When a Product is Required to be Maintained under Refrigeration.

(L&R, 1990, p. 86)

Background

Section 6.5.(b) was revised to clarify that the reference temperature of 4.4 °C (40 °F) applies only to products that must be refrigerated to maintain product quality, rather than to items, such as carbonated soft drinks, that are refrigerated for the purchaser's convenience.

Guideline

The Committee also discussed how an inspector could decide whether a product under refrigeration is required to be maintained under refrigeration. The following guidelines are provided:

1. The traditional food items that normally require refrigeration and are found in refrigerated cases will not ordinarily have any statement about requiring refrigeration. These items include milk, orange juice, and similar products. They may be tested at any temperature at, above or below their reference temperature of 40 °F (4 °C) because such products are at their maximum density at their reference temperature, and the volume of such products will always increase at higher or lower temperatures. Thus any errors made by not measuring at the exact reference temperature will be in the favor of the packer.

2. Food items that normally require refrigeration, but which are processed so as not to require refrigeration prior to opening, will have "refrigerate after opening" or similar wording on the label. Such items as milk and orange juice can be found in this category as well as in the "refrigeration required" category. The two categories can be distinguished by the "refrigerate after opening" statement, which calls for testing at or above their reference temperature of 68 °F (20 °C).

3. Food items that are not expected to require refrigeration, but which may be refrigerated for the convenience of the consumer (such as carbonated beverages), are to be tested at temperatures of 68 °F (20 °C) or above even when found refrigerated for the convenience of the consumer.

2.2.13. Declaration of Identity: Consumer Package (UPLR) and 1.5.1. in Combination with Other Foods (UMSCR).

(L&R, 1990, p. 93)

Background

Many food products are made by the retail store and labeled with names that may or may not have standards of identity or standards of composition in federal regulation or policy (for example, chicken cordon bleu). Weights and measures officials need to know which names have standards of identity that must be followed in formulating the product and, therefore, in providing the ingredient statement.

Food Standards

The U.S. Department of Agriculture's Food Safety and Inspection Service (USDA - FSIS) and the U.S. Department of Health and Human Services' FDA share the responsibility of assuring truthful and accurate information on product labels. USDA - FSIS has responsibility for the development and application of the labeling requirements applicable to meat and poultry products containing more than 3 % fresh meat or at least 2 % cooked poultry meat. FDA oversees the labeling of most other food products.

USDA Standards of Identity and Composition

USDA has statutory authority to establish standards of identity for meat and poultry products. A standard of identity prescribes a manner of preparation and the ingredients of a product that is labeled with a particular name. A food that bears the name of a standardized food that does not satisfy the requirements of the applicable standard is misbranded. Examples of standardized products include: "Ham," "Ham Water Added," "Hot Dogs," "Chicken and Noodles," and "Spaghetti Sauce with Meatballs."

Almost all standards enforced by FSIS are called "standards of composition." These standards identify the minimum amount of meat or poultry required in a product's recipe. For example, the standard of composition for "beef a la king" states that, if a product carries this name on its label, at least 20 % cooked beef must be used in the recipe.

But standards of composition do not prevent a manufacturer from increasing the meat or poultry content or adding other ingredients to increase a product's appeal. For instance, a processor has the option of using more than the required amount of beef in beef a la king and adding other ingredients to make the product unique. A listing of meat and poultry content and labeling requirement including terms that are further defined can be found in the USDA FSIS Food Standards and Labeling Policy book which is available at **www.fsis.usda.gov/OPPDE/larc/Policies/Labeling_Policy_book_082005.pdf**.

Label Approval

Food manufacturers are responsible for compliance with the FSIS labeling rules and adherence to the process maintained by FSIS for the evaluation and approval of meat and poultry product labels. This Guide provides the basic information necessary to devise a label for meat and poultry products and to understand the regulatory process administered by FSIS. A Guide to Federal Food Labeling Requirements for Meat and Poultry Products (2007) URL is located at **www.fsis.usda.gov/PDF/Labeling_Requirements_Guide.pdf**.

2.2.14. Typewriter and Computer Printer Ribbons and Tapes.

(L&R, 1991)

Interpretation

Typewriter and computer printer ribbons must be labeled by length. In addition, character yield information may be disclosed on the principal display panel.

Background

Packages of typewriter and computer printer ribbons and tapes have been found in the marketplace with no declaration of quantity of any kind. There is information on the package about the type of machine the ribbon or tape is designed to fit, but this is not a declaration of quantity. Purchasers have been misled as a result of the failure of some manufacturers to disclose the length; ribbons designated for a particular machine may be sold at a low price, but with substantially less length than ribbons ordinarily produced for the machine.

2.3.1. Instant Concentrated Products.
(L&R, 1977, p. 219)

Interpretation

No additional net contents information (other than weight) is required for instant coffee, tea, and cocoa.

Background

It was proposed that certain products, such as instant coffee, tea, and cocoa, should have a dual statement of weight including the number of cups (e.g., makes ten 6 oz cups).

The National Coffee Association of U.S.A., Inc., offered the following comments:

1. The number of servings of instant coffee will depend upon the size of the cup involved and the taste of the individual consumer.

2. The size of a cup will vary widely, ranging from a small "demitasse" cup to a large coffee mug.

3. The taste of the individual consumer defies definition because it will vary as widely as the number of individuals considered. Market research shows many like it "strong and black" and others prefer it "mild and thin."

4. Any statement placed on a container of instant coffee that represents that the consumer will be able to obtain a specified number of servings would be arbitrary, confusing and, in a very sense, deceptive.

5. In view of the foregoing, any such requirements that the number of servings be listed on a container of instant coffee might expose the manufacturer to complaints from consumers that it was engaging in an unfair and deceptive practice.

Other issues that the Committee discussed included the authority to require precise directions (rather than, for example, 2 to 3 heaping teaspoons) and the issues of product variability and uniform enforcement.

2.3.2. Fresh Fruits and Vegetables.
(L&R, 1979, p. 176; 1980; 1982, p. 152; 2008)

Guideline

Recognizing the difficulty faced by consumers when more than one method of sale is employed in the same outlet for the same product, non-comparable methods of sale (e.g., weight and measure) for the same produce item in the same outlet should be minimized.

This guideline applies to all sales of fruits and vegetables. There are two tables, one for specific commodities and one for general commodity groups. Search the specific list first to find those commodities that either do not fit into any of the general groups or have unique methods of sale. If the item is not listed, find the general group in the second table. The item may be sold by any method of sale marked with an X.
(Amended 2008)

Method of Retail Sale for Fresh Fruits and Vegetables Specific Commodity					
Commodity	**Weight**	**Count**	**Head or Bunch**	**Dry Measure (any size)**	**Dry Measure (1 dry qt or larger)**
Artichokes	X	X			
Asparagus	X		X		
Avocados		X			
Bananas	X	X			
Beans (green, yellow, etc.)	X				X
Brussels Sprouts (loose)	X				
Brussels Sprouts (on stalk)			X		
Cherries	X			X	X
Coconuts	X	X			
Corn on the Cob		X			X
Dates	X				
Eggplant	X	X			
Figs	X				
Grapes	X				
Melons (cut in pieces)	X				
Mushrooms (small)	X			X	X
Mushrooms (portobello, large)	X	X			
Okra	X				
Peas	X				X
Peppers (bell and other varieties)	X	X			X
Pineapples	X	X			
Rhubarb	X		X		
Tomatoes (except cherry/grape)	X	X			X

Method of Retail Sale for Fresh Fruits and Vegetables General Commodity Groups					
Commodity	**Weight**	**Count**	**Head or Bunch**	**Dry Measure (any size)**	**Dry Measure (1 dry qt or larger)**
Berries and Cherry/Grape Tomatoes	X			X	
Citrus Fruits (oranges, grapefruits, lemons, etc.)	X	X			X
Edible Bulbs (onions [spring or green], garlic, leeks, etc.)	X	X	X		X
Edible Tubers (Irish potatoes, sweet potatoes, ginger, horseradish, etc.)	X				X
Flower Vegetables (broccoli, cauliflower, brussel sprouts, etc.)	X		X		
Gourd Vegetables (cucumbers, squash, melons, etc.)	X	X			X
Leaf Vegetables (lettuce, cabbage, celery, etc.)	X		X		
Leaf Vegetables (parsley, herbs, loose greens)	X		X	X	
Pitted Fruits (peaches, plums, prunes, etc.)	X	X			X
Pome Fruits (apples, pears, mangoes, etc.)	X	X			X
Root Vegetables (turnips, carrots, radishes, etc.)	X		X		

2.3.3. Cardboard Cartons.
(L&R, 1974, p. 223)

Guidelines and Interpretations

Cardboard cartons should be sold by their dimensions. Identification numbers used in the trade do not correspond to these dimensions and could tend to mislead the uninformed purchaser (although there is no actual unit such as inches associated with the identification numbers). Sales or catalogue literature will have to be investigated to determine whether there is sufficient information upon which to make a purchasing decision.

Background

Copies of letters received by the New York Bureau of Weights and Measures regarding cardboard containers were forwarded to the Committee. These letters highlight the confusion that exists when these containers are sold to new businessmen by an identity number which is often mistaken for the size of the box. For example, a 30 x 4 identification number refers to a box whose actual size is 27 x 3 inches. It was suggested that a new section be added to the Method of Sale of Commodities Regulation so that these containers can be sold on a basis that will provide more accurate information.

An important argument in support of adding a new section is that small businessmen just getting started need as much assistance as can be provided in order to survive and grow.

An argument opposing this change is that a table, similar to Table 1. of Section 2.9. (Softwood Lumber) of the Uniform Method of Sale Regulation, could be printed showing the relationship between identity and size; this would not solve the problem.

It is the consensus of the Committee that these containers should be sold by actual size. The Committee does not believe, however, that every trade practice must be controlled through the Uniform Laws and Regulations. This is particularly true where the item does not directly concern the retail consumer. The Committee, therefore, recommends that the appropriate trade associations be contacted and asked to correct this practice on a voluntary basis.

2.3.4. Catalyst Beads.

(L&R, 1981, p. 100)

Guideline and Interpretation

The proper method of sale of catalyst beads used in automobile exhaust systems is by volume. It is appropriate for the quantity declaration to be supplemented by part number or other description of the specific converter for which the package of catalyst beads is intended.

Background

A communication from the General Motors Corporation AC Spark Plug Division was forwarded to the Committee which proposes discontinuing the labeling of their catalyst beads by weight. When the catalyst becomes contaminated by leaded gasoline or prolonged use, the catalytic converter in the exhaust system of recent GM cars and trucks (running on unleaded gasoline) must be emptied of its catalyst beads and be refilled by volume with replacement catalyst beads in order to meet emission standards. The beads are used by volume (to fill a catalytic converter), are hygroscopic, and vary in core material density. Therefore, packages of beads meeting a net weight label require an additional one-third pound (on the average) over the packages labeled by volume, cost about $7.50 more per package, and the additional weight of beads will be discarded in actual use.

2.3.5. Incense.

(L&R, 1978, p. 151)

Interpretation

Incense labeled by count is fully informative and sufficient.

Background

The State of Oregon raised the issue of proper quantity declarations for the sale of incense. The question is what, if any, information other than count, such as weight or volume or length, is necessary for an adequate description on packages of incense. The Committee is of the opinion that a statement of count as defined in Section 6.4.1(c) of the Uniform Packaging and Labeling Regulation is fully informative and is sufficient in this case.

2.3.6. Sea Shells.

(L&R, 1976, p. 223)

Guideline

Sea shells shall be sold by count and weight for packages of 50 sea shells or less and by volume and weight for packages containing more than 50 sea shells.

2.3.7. Tire Tread Rubber Products.

(L&R, 1976, p. 233)

Guideline

Tire tread rubber products shall be sold by net weight. The polyethylene film protective backing shall be part of the product and included in the net weight. The core is part of the tare and must be deducted from the gross weight to determine the net weight.

2.3.8. Wiper Blades.
(L&R, 1979, p. 182)

Interpretation

There is a trade custom of labeling automobile wiper blades by the length of the metal backing or vertebra, not the length of the blade. This is an acceptable method of sale and net contents declaration.

Background

The Committee received a request from a manufacturer of automobile wiper blades that had a problem with one state concerning the measurement of length as labeled on their packages. The state felt that the proper designation should be the length of the blade itself; the manufacturer said that traditionally the industry measured the length of the metal backing or vertebra.

The Committee, after some discussion, determined that since there was no intent to mislead customers, the traditional measurement of the metal backing or vertebra was acceptable.

2.3.9. Fireplace Logs.
(L&R, 1975, p. 174)

Interpretation

Time of burning is not an appropriate quantity declaration for fireplace logs. (Section 2.4.3. of the Uniform Method of Sale of Commodities requires single logs to be sold by weight, or if packaged and less than 4 ft^3, weight plus count.)

Background

The enforceability of quantity declarations using time as the basis of measurement for commodities, including packaged commodities, must be considered carefully if equity in the marketplace is to be achieved. The Committee wants to stress to those who have submitted time declaration questions that the enforceability factor should not override consumer protection and uniformity considerations. Based on the above criteria, the Committee recommends that the Conference take the position that time is not an appropriate quantity declaration for fireplace logs.

2.3.11. Packaged Foods or Cosmetics Sold from Vending Machines.
(L&R, 1982, p. 152)

Interpretation

Packaged foods and cosmetics sold from vending machines must be labeled the same as similar items not sold in vending machines, including identity, responsibility, net contents, and ingredient declaration, except that Section 3.3. of the Uniform Regulation for the Method of Sale of Commodities permits identity and net contents to be posted on the machine in lieu of appearing on the package.

Background

As part of its review of the Uniform Regulation for the Method of Sale of Commodities, the FDA recommended adding a statement to Section 3.3. that packaged foods and cosmetics sold in vending machines must in general be labeled in accordance with requirements for similar articles not sold in vending machines (e.g., ingredient declaration requirements). The Committee recommends that this information be made a guideline rather than incorporated as part of the uniform regulation.

2.3.12. Movie Films, Tapes, Cassettes.

(L&R, 1975, p. 174)

Guideline

Movie film may be sold by linear measure. Magnetic tapes and cassettes may be sold by either linear measure or playing time.

Background

The enforceability of quantity declarations using time as the basis of measurement for commodities, including packaged commodities, must be carefully considered to achieve equity in the marketplace. The Committee wants to stress to those who have submitted time declaration questions that the enforceability factor should not override consumer protection and uniformity considerations. The committee further recommends that the states follow FTC guidelines in requiring lineal measure for the sale of movie films and permit either linear measure or playing time for magnetic tapes and cassettes.

2.3.13. Vegetable Oil.

(L&R, 1983, p. 208)

Guideline and Interpretation

Packaged liquid vegetable oil must be labeled by liquid volume, although net weight may also be declared.

Background

Packages of liquid vegetable oil are being sold for restaurant and other small food business use labeled by weight. It has been brought to the attention of the Committee that containers of product labeled "5 gal" look identical in dimensions to those labeled "35 lb" but the density of the vegetable oil is such that the 35 lb cans contain only about 4½ gal. The Institute of Shortening and Edible Oils indicated that companies selling liquid vegetable oils often compete with those selling solid shortening, and that a net weight comparison is useful for these purposes. Recipes for food products in large sizes sometimes provide ingredient quantities by weight or volume.

It is the opinion of the members of the Committee that packaged liquid vegetable oil must be labeled by liquid volume although a net weight may be declared in addition to the net volume statement.

When a single manufacturer of vegetable oil packages the same oil in the same size container with two such widely different net quantity statements, this practice could easily be considered (a) misleading to the customer, and (b) nonfunctional slack fill. Weights and measures enforcement action should be taken.

2.3.15. Bulk Sales.

(L&R Committee, 1986, p. 140)

When packaged or wrapped items (such as individually wrapped candies) are sold from bulk displays by weight, the price must be based on the net weight, not the weight including the individual piece wrappings. This will require (1) subtracting the weight of the bag into which the customer puts the pieces plus (2) subtracting the weight of the piece wrappings (the latter is a percentage of the gross weight – that is, the tare increases as the customer selects more of the commodity).

Background

Retail food stores are merchandising prepackaged commodities such as candies, pet food, snack bars, and bouillon cubes from bulk displays. Some retailers sell these products by gross weight. Section 1.2. of the Uniform Weights and Measures Law reads in part: "The term 'weight' as used in connection with any commodity means net weight. . ."

A workshop was held on June 20, 1986, at the U.S. Department of Commerce, Washington, D.C., to explore the issues and alternatives involved in the sale of prepackaged goods from the bulk food sales areas of supermarkets. Representatives of the packaging, supermarket, and small grocery industries, scale and point-of-sale (POS) systems manufacturers, the U.S. Food and Drug Administration, weights and measures agencies, and the National Institute of Standards and Technology attended. No final recommendations came from this meeting; however, the participants expressed an interest in meeting again after a written report of the June 20, 1986, meeting was made available and before the Interim Meetings of the NCWM in January 1987. The following issues were discussed:

1. Prepackaged commodities in bulk displays are being sold on a gross weight basis.

   Federal regulations covering packaged goods and every state Weights and Measures Law require any sale by weight to be "net weight" (not including the weight of the wrapping materials). In some areas of the nation, many items are being sold on a gross weight basis in the supermarkets, for example, fresh fruit and vegetables in poly bags in the produce area. Perhaps because of the light weight of these bags (that is, the minimum size of the scale division on the ordinary supermarket checkout scale is large with respect to the weight of the poly bags), low priority is given to correcting this sales practice, and a lack of uniformity in enforcement of the net weight requirements results. Weights and measures officials have found tare amounting to over 40 % of the gross weight in prepackaged items sold from bulk; the majority of cases seems to range from 3 % to 12 %. Officials see the need to "draw the line" in a sales practice that appears to have evolved from other practices that were not heavily monitored and corrected at their inception.

2. Retailers face technical and administrative problems in properly deducting tare from the gross weight.

   Automatic deduction of tare is preferable for large-scale retailers because of its speed. No equipment (either stand-alone scale or POS) is available at the present time that can: (1) subtract a percentage of the gross weight to represent the tare weight; or (2) subtract a fixed tare for the bag and a percentage tare for the wrapper on the prepackaged item. [Editor's Note: There is equipment now available that can deduct a tare that is a percentage of the gross weight.] Two POS system manufacturers said that new systems with percentage tare capability could be designed, but they could not definitely say whether retrofitting existing systems was possible. They said that the ability to retrofit declined with the age of the system. Supermarket representatives expressed concern that their in-store computer software would need modification above and beyond the retrofitting or software redesign that might be done by the POS manufacturers; their software is designed around current POS software.

   Deduction of tare in the bulk food area using a scale other than the checkout scale can be done more easily than at checkout if a POS system is being used. A tare look-up table used in conjunction with the scale appears to be the only currently used method that meets the net weight requirements when packaged products are sold from bulk. (The procedure is to gross weigh the product, look up the tare, subtract it from the gross weight, and then determine a final net weight and total price.)

   Each retailer will have to consider the cost of additional manpower (as the weighing and marking of the purchase in the bulk food area might require), new equipment (purchasing scales or POS systems with percentage tare capability), or retrofit of existing equipment as compared with the value of the market share contributed by the bulk marketing of prepacked commodities. However, two supermarket chain representatives said that they expected some growth in this type of sale (because of the customers' perception of cleanliness of the product, for example).

3. Present methods of sale and advertising are often misleading.

   Suggestions were made that advertising on a "wrapped weight" basis would properly inform the consumer. However, it was pointed out that a typical purchaser does not know what "wrapped weight" is (i.e., gross weight). Moreover, selling packaged goods on a gross weight basis is illegal; it thwarts value comparison with other products sold by net weight.

Bulk food sales advertising often includes claims of savings of, for example, 10 % to 20 % over a purchase of the same commodity in standard-pack form. These advertising claims can be exaggerated and misleading if the comparisons referenced are between standard-pack commodities sold net weight and products sold from bulk on a gross weight basis.

The possibility of advertising a net weight unit price, but actually weighing at the checkout on a gross weight basis (and charging at a lower gross weight unit price) was discussed. For example, a sign could be posted with the following:

> "$1.50 per pound, net weight. We are not able to weigh this packaged product on a net weight basis (that is, without the wrapper), and will therefore charge you $1.40 per pound including the wrapper weight at the checkout."

Everyone agreed that advertising claims and appropriate wording would have to be chosen carefully if this is to be viable. However, those weights and measures officials present were generally opposed to this alternative based on the difficulty of enforcement and lack of assurance that a consumer would really understand explanatory signage.

2.5.6. Guidelines for NCWM Resolution of Requests for Recognition of Moisture Loss in Other Packaged Products.

(Exec, 1988, p. 94)

The Task Force on Commodity Requirements limited its work to only a few product categories, using these categories as models for addressing moisture loss. The gray-area concept is the result of this work.

Recognizing several candidates for future work in moisture loss, the Task Force recommends that the following guidelines for moisture loss be followed as far as possible by any industry requesting consideration:

1. There should be reasonable uniformity in the moisture content of the product category. For example, since pet food has final moisture contents ranging from very moist to very dry, some sub categorization of pet food needs to be defined by industry before the NCWM study of the issue.

2. The predominant type of moisture loss (whether into the atmosphere or into the packaging materials) must be specified.

3. Different types of packaging might make it necessary to subcategorize the product. For example, pasta is packaged in cardboard, in polyethylene, or other packaging more impervious to moisture loss. The industry should define the domain of packaging materials to be considered.

4. "Real-world" data is needed on the product as found in the retail marketing chain – not just laboratory moisture-loss data.

5. The industry requesting consideration of moisture loss for its product should collect data on an industry-wide basis (rather than from only one or two companies).

   Information concerning the relative fractions of imported and domestically produced product should be available, for example, in order to assess the feasibility of interacting with the manufacturer on specific problem lots.

6. Moisture loss may occur either:

   - during manufacturing; or
   - during distribution.

Data will be needed to show the relative proportion of moisture loss in these different locations since moisture loss is permitted only under good distribution practices. Geographical and seasonal variations may apply.

7. A description of the processing and packaging methods in use in the industry will be of great value, as will a description of the distribution system and time for manufacturing and distribution. A description of the existing net quantity control programs in place should be given, together with information on how compliance with Handbook 133 is obtained. A description of maintenance and inspection procedures for the scales should be provided, together with information on suitability of equipment and other measurements under Handbook 44.

8. A description of federal and local agency jurisdiction and test should be given, as well as any regulatory history with respect to moisture loss and short weight. Has weights and measures enforcement generated the request? What efforts have addressed the moisture loss issue prior to approaching the NCWM? Are the appropriate federal agencies aware of the industry's request to the NCWM?

9. The industry should propose the type of compliance system and/or moisture determination methodology to be used. The compliance scheme, if it contains industry data components, should be susceptible to verification (examples: USDA net weight tests for meat; exchange of samples with millers for flour) and should state what the companies will do to provide data to field inspection agencies in an ongoing fashion (as the gray-area approach requires). If in-plant testing is to be combined with field testing, who is to do such testing, and how is this to be accomplished? It should be possible to incorporate the proposed testing scheme into Handbook 133 to be used with Category A or B sampling plans.

When all the preliminary information recommended above has been collected, a field test of the proposed compliance scheme should be conducted by weights and measures enforcement officials to prove its viability.

See the plan diagrammed on the next page.

Plan For NCWM Resolution of Individual Requests For Recognition of Moisture Loss

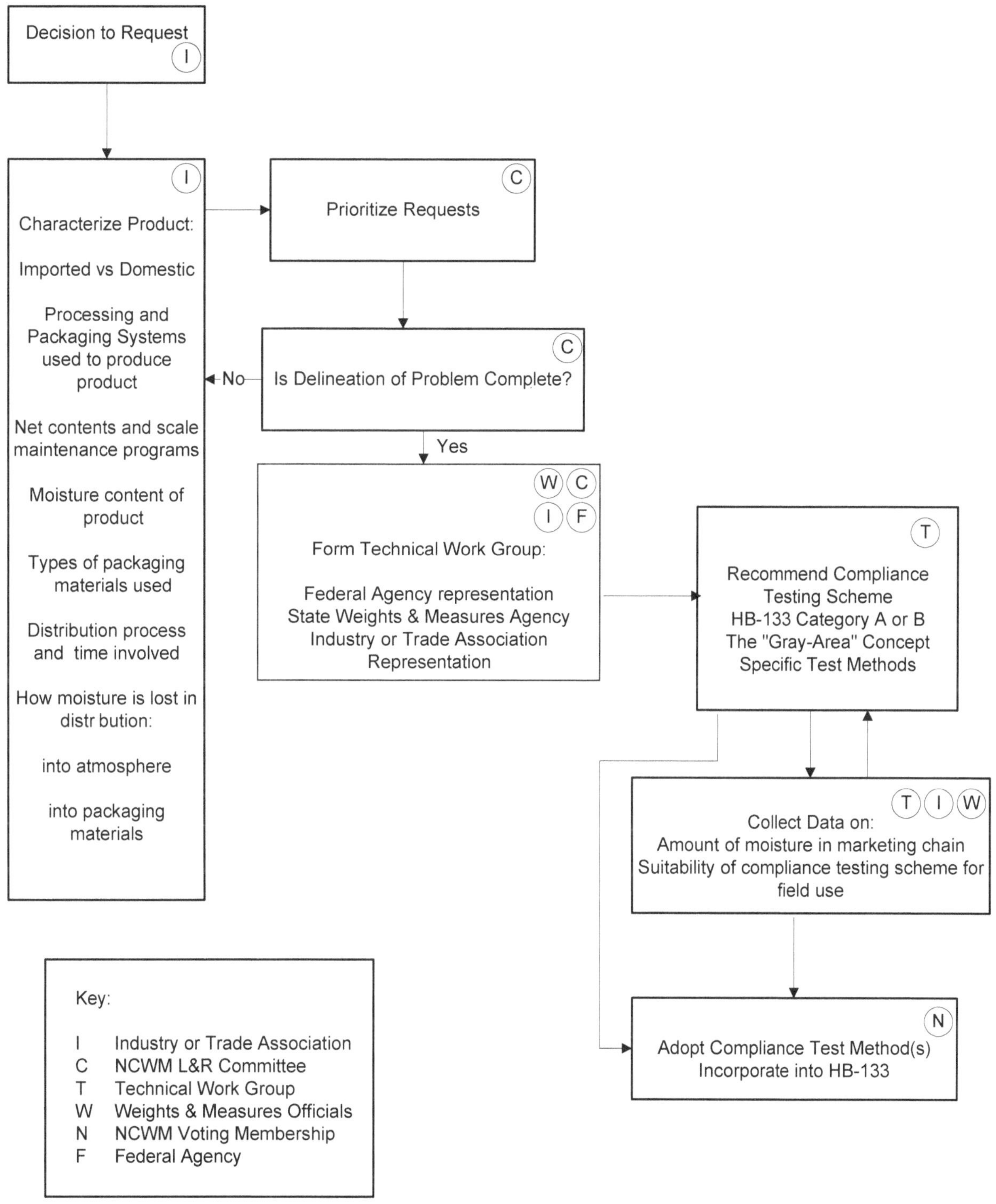

2.6.1. Retail Gas Sales and Metric Price Computations in General.
(S&T, 1980, p. 227)

Guideline

The National Institute of Standards and Technology published equivalent rounded values for metric equivalents of inch-pound units should be used. They are:

3.785 411 784 liters = 1 gallon
0.264 172 052 4 gallon = 1 liter

A "Rule of Reason" should apply to the corrected value so that the value used is consistent with the quantity of the transaction. The converted value should never have fewer than four significant digits and should have at least the same number of significant digits as the number of significant digits in the quantity of product being converted. For example, if a 1000 gal delivery were to be converted to liters the value would be 3785 L; for 10 000 gal, 37 854 L; for 100 gal, 378.5 L.

In the case of expressing a unit price equivalent for consumer value comparisons in retail gasoline sales, the following formula should be used: (advertised, posted, or computing device unit price per liter) x 3.785 = (equivalent unit price per gallon, rounded to the nearest $^{1}/_{10}$ cent.)

Examples:
26.9 cents per liter x 3.785 = $1.018 per gallon
26.8 cents per liter x 3.785 = $1.014 per gallon
26.7 cents per liter x 3.785 = $1.011 per gallon
26.5 cents per liter x 3.785 = $1.003 per gallon
26.4 cents per liter x 3.785 = $0.999 per gallon

This method is preferable to the alternative method of dividing the price per gallon by 3.785, which results in the same price per liter for three or more different prices per gallon when rounded to the $^{1}/_{10}$ cent.

2.6.2. Price Posting.
(L&R, 1981, p. 101)

Guideline

1. Street Signs.

   a. Until such time as the sale of gasoline and other Engine fuels is predominately by metric measurement (liter), price per gallon information should be made readily available to all prospective customers.

   b. All street, roadside, and similar advertising signs displaying product price should provide price per gallon information.

   c. Signs showing the equivalent price per liter may also be used, but their use is optional and should not employ numerals larger than the equivalent gallon price display.

   d. Signs should show complete dollar and cents numerals, and they should be clearly legible and of full size. An exception should be granted to street signs that were designed to display only three numerals (e.g., $.899) and not four numerals as required for prices over $1.00 per gallon (e.g., $1.259). Until such signs can be replaced or modified, it would be acceptable:

      (1) to attach an appropriate sign extension with the decimal fraction of a cent representation in alignment with the posted price;

(2) to include a smaller fraction of a cent representation with the last numeral of the posted price; or

(3) to add the whole number "one" before the cents values.

e. The changeover to advertising prices by the liter as a single mode of pricing should be established when 75 % of all retail outlets in a jurisdiction have converted their dispensers to metric measurement.

2. Posting of Prices at the Dispenser.

Each retail outlet should use exclusively only one measurement method of sale (gallon or liter). A change from one method to another should be carried out for all devices dispensing motor fuels in the retail outlet.

In the case of liter sales, suitable posting of per gallon and per liter prices at the device, service island, premises of the retail outlet, or any other locations must be in accordance with state and local laws, regulations, and ordinances, and in a manner that facilitates consumer comparisons between the per gallon price and the per liter price. Additional requirements may be necessary to avoid uncertainty as to nomenclature, location, and size of information on signs.

It is recommended that:

a. Current and accurate price comparisons between gallon and liter values be posted at the dispenser within easy view of the customer and visible from either side of the island.

b. The sign should show equivalent quantity and price information.

Examples:
27.1¢ per liter = $1.026 per gallon
3.785 liters = 1 gallon

c. Letters and numerals should be at least ¾ in (19 mm) in height and ⅛ in (3 mm) in width of stroke.

3. Quantity and Price Display on Dispensers

It is required that dispensers be designed to clearly show all required quantity and price information on the face(s) of a motor fuel dispenser in accordance with Handbook 44.

4. Dispenser Modification Kits

As an interim alternative to "half pricing," a number of computer modification kits have been installed to modify existing retail motor fuel dispensers that were not designed to compute and indicate prices over 99.9¢ per gallon.

Some of the modification kits that have been referred to state weights and measures officials for approval have been rejected as failing to conform to Handbook 44 requirements. It is recommended that all modification kits and future modifications of dispensers be so designed and made as to be in full compliance with all applicable requirements of Handbook 44.

2.6.3. Octane Posting Regulations.
(Liaison, 1979, p. 240)

Guideline

Weights and Measures officials should report to the FTC any instances of failure to post octane ratings by service stations. These would most likely occur during routine inspections of service station gasoline dispensers. Reports should be made to the appropriate FTC regional offices as listed below.

Background

As of June 1, 1979, the FTC requires the determination of octane ratings by refiners, the certification of octane ratings by refiners and distributors, and the posting of octane ratings by retailers on all gasoline pumps. The requirements are set forth in Public Law 95 297, the Petroleum Marketing Practices Act (PMPA), passed in June 1978 and the FTC's Octane Rule, 16.C.F.R. Part 306. Although the octane posting rule has no effect on most FTC programs administered by state weights and measures officials with respect to checking gasoline dispensing devices for accuracy, the Liaison Committee feels that the Conference should be generally informed about the law and the FTC rule, if only to be prepared to answer inquiries about it or for some possible future enforcement demands. Keeping apprised of developments associated with the rule may be advisable. In addition, it will affect states which have octane certification and posting programs.

Regional Offices, Addresses, and Telephone Numbers:

Northeast Region (CT, ME, MA, NH, NJ, NY, PR, RI, VT, and U.S. Virgin Islands) Federal Trade Commission One Bowling Green New York, NY 10004 (877) 382-4357	**East Central Region** (DE, DC, MD, MI, OH, PA, VA, and WV) Federal Trade Commission 1111 Superior Avenue Suite 200 Cleveland, OH 44114-2507 (877) 382-4357	**Southeast Region** (AL, FL, GA, MS, NC, SC, and TN) Federal Trade Commission Suite 1500 225 Peachtree Street, NE Atlanta, GA 30303 (877) 382-4357
Midwest Region (IL, IN, IA, KS, KY, NE, ND, MN, MO, SD, and WI) Federal Trade Commission 55 West Monroe Street Suite 1825 Chicago, IL 60603 (877) 382-4357	**Northwest Region** (AK, ID, MT, OR, WA, WY) Federal Trade Commission 915 Second Avenue, Room 2896 Seattle, WA 98174 (877) 382-4357	**Southwest Region** (AR, LA, NM, OK, and TX) Federal Trade Commission 1999 Bryan Street Suite 2150 Dallas, TX 75201-6808 (877) 382-4357
Western Region (AZ, Northern CA, Southern CA, CO, HI, NV, and UT) Federal Trade Commission 901 Market Street Suite 570 San Francisco, CA 94103 (877) 382-4357	**Western Region** (AZ, Northern CA, Southern CA, CO, HI, NV, and UT) Federal Trade Commission 10877 Wilshire Boulevard Suite 700 Los Angeles, CA 90024 (877) 382-4357	

The preemption section of PMPA (204) reads as follows:

Section 204. To the extent that any provision of this title applies to any act or omission, no state or any political subdivision thereof may adopt, enforce, or continue in effect any provision of any law or regulation (including any remedy or penalty applicable to any violation thereof) with respect to such act or omission, unless such provision of such law or regulation is the same as the applicable provision of this title.

Section 204 prohibits states and other political subdivisions from enforcing requirements that are not the same as the applicable provisions of this law. Jurisdictions having octane requirements should carefully review with their legal advisors the effect of this law.

The FTC's Octane rule was published in final form on March 30, 1979, in the Federal Register (Vol. 44, No. 63, Part V, pp. 19160 19172). The rule became effective June 1, 1979.

The law requires that refiners determine octane ratings of their products, and certify them to their distributors. The distributors must pass along the certification to the retailer, unless he blends the gas, in which case he may have to certify his blend.

A similar procedure relating to the posting of octane ratings is set forth for the retailer. The FTC is responsible for enforcement with respect to the accuracy of the certified ratings. The FTC is also empowered to check records, which must be retained for one year by each link in the distribution chain.

The FTC is in need of help from the state and local jurisdictions in the area of surveillance and testing. Such assistance could occur at a number of levels. Notice of octane mislabeling and failure to post octane ratings is requested.

Other levels of assistance would concern jurisdictions that have octane testing programs and would be interested in cooperating with FTC in testing or in reporting discrepancies in octane rating.

For more information contact the Federal Trade Commission at 600 Pennsylvania Avenue, NW, Washington, D.C. 20580, phone (202) 326-2222.

2.6.4. Multi-Tier Pricing: Motor Fuel Deliveries (Computing Pumps or Dispensers).
(L&R, 1982, p. 150; L&R, 1985, p. 100) (L&R, 1988, p. 162)

Policy

Charging different prices for the same product depending upon the manner of payment, other purchases, amount of service, etc., is a management decision of the merchandiser. Those merchants who elect to offer multiple prices for motor fuel must comply with the state and local weights and measure laws and regulations, including Handbook 44. They must also make marketing decisions that comply with state truth in lending, cash discount, price advertising, and usury laws. All such laws are intended to prohibit deceptive, misleading, or misrepresentative information being given to the consumer. The following guidelines are intended to apply to price advertising or posting at the street side or highway as well as at the pump or dispenser, and to the price computed at the device. These guidelines are applicable to other discount or combination offers (such as combination purchases of car wash and gas, for example).

1. If a price is posted or advertised, it must be available to all qualified customers. If any condition or qualification is required to obtain the posted price, that condition must also be posted clearly and understandably, in conjunction with the price wherever it is posted.

2. The lowest price may be posted or advertised by itself as long as any restrictions for receiving that price (for example, "cash only") are also clearly posted or advertised in conjunction with the price and as long as other state requirements do not prohibit it. For example, certain states require that all prices available from a given retail location must be posted on street side signs if any prices are posted.

3. If the merchandiser elects to establish separate devices or islands for sale of the same product at different prices, the devices or islands shall be clearly identified as "cash," "credit," "self-serve," or other appropriate wording to avoid customer confusion.

4. The use of a single-price-computing dispenser for sale of motor fuel at multiple unit prices is inappropriate, facilitates fraud, and should be eliminated. The NCWM should adopt a plan and timetable for changeover to devices that can compute and display final money values for multiple prices.

2.6.5. Cereal Grains and Oil Seeds.

(L&R, 1981, p. 95; L&R, 1996, p. 135)

Interpretation

The addition of water to grain for the purpose of adding weight prior to selling grain by weight is an illegal practice under federal laws.

NOTE: *Effective February 11, 1995, the Federal Grain Inspection Service adopted a regulation in 7 CFR Part 800.61 prohibiting the application of water to grain except for milling, malting, or similar processing operations. See Volume 59, No. 198 for Friday, October 14, 1994, or page 52 071, for additional information.*

Background

A letter from the Oklahoma Grain and Feed Association was forwarded to the Committee asking whether the addition of water to grain is legal. The request was prompted by an article reporting on methods of adding water to grain to bring the moisture content up to market standards. For example, when soybeans are sold at 8 % moisture content, there is less weight sold (and less revenue for the soybeans to the seller) than if water were added to the same soybeans to bring them to 10 % moisture content.

However, the Committee is greatly concerned about the ramifications of such practices. Many grain experts do not believe that over-dried grain should be valued as highly as grain at moisture contents close to market standards. Overly dry grain is more susceptible to breakage, for example.

Water added after harvest will not be taken up chemically the way that naturally moist grain binds water. Errors in adding water or the particular biochemical nature of the grain after addition of water can lead to spoiled grain. Studies on the long term keeping qualities of grain with water added have not been carried out. The calibration of moisture meters is based on naturally moist grain, and there is a known difference between the electrical properties of naturally moist grain and grain with moisture added.

Of a more basic nature, however, the Committee recognizes the fact that a grain buyer purchases grain expecting such grain to be naturally moist or dried, not to be with water added. The seller who adds water to grain solely to add weight, therefore, misrepresents his product.

Both the FDA and USDA have sent letters to the Committee indicating that the addition of water to grain solely for the purpose of adding weight is an illegal practice. Because existing federal laws already prohibit this practice, the Committee recommends no further action on the part of the Conference at this time.

2.6.6. Basic Engine Fuels, Petroleum Products, and Lubricants Laboratory.

(L&R, 1994, p. 129-135; L&R, 2006, p. L&R-8) (Developed by the Petroleum Subcommittee.)

The petroleum fuels and lubricants laboratory is an integral element of an inspection program and is generally developed to satisfy the testing requirements as described in the laws and rules of the regulating agency. Guidelines have been developed to assist states in evaluating their options of employing a private lab or building or expanding their own lab. This information is available at **www.nist.gov/pml/wmd/index.cfm**.

2.6.7. Product Conformance Statements.

(L&R, 1992, p. 148)

Interpretation

References to a product's conformance with product standards (for example, "manufactured to standard EN235" or similar product conformance statements) on labels for wallcovering or other products, are not considered qualifying terms and do not violate Section 6.12.1., Supplementary Quantity Declarations of the Uniform Packaging and Labeling Regulation, provided the requirements of Section 8.1.4. Free Area is met.

Background

The Wallcovering Manufacturers Association (WMA) requested the Conference's position on the use of conformance statements on the labels of wallcovering and border material. This issue relates to wallcovering products that originate from manufacturers in Europe where a declaration of conformance to a specific government standard is required on consumer packages. Thousands of product "standards" or "Euronorms" are being established for the European Community. Conformance declarations are required to provide consumers and customs officials with information on the product. The issue relates to the use of such statements as "manufactured to standard EN235" on labels of wallcovering that are imported from Europe. The WMA requested the Committee's opinion on the use of this type of statement if a package is labeled in conformance with sections Section 6.12.1. Supplementary Quantity Declarations and Section 8.1.4. Free Area. One question is whether the display of the conformance statement would be permitted provided that it did not include an unacceptable quantity declaration. Another question concerns the need to comply with the requirement for adequate free area around the quantity declaration when the conformance declaration is placed on the label. It was the Committee's opinion that conformance statements on package labels would not violate any provisions of the PLR if the requirements of Sections 6.12.1. and 8.1.4. are met.

The Committee recommended this interpretation for inclusion in Handbook 130 because it is likely that this type of notice will become common as more and more free market trading areas are opened to expand international trade. This interpretation does not indicate acceptance or endorsement of any requirements contained in product conformance statements.

2.6.8. Commodities Under FTC Jurisdiction under the Fair Packaging and Labeling Act and Exclusions.

(L&R, 1993, p. 279; L&R, 1994, p. 294)

The following lists indicate the commodities and commodity groups that are and are not within the scope of the Fair Packaging and Labeling Act administered by the FTC. The following codes appear with each excluded commodity and designate the reason that the particular commodity has been excluded.

BATF – designates commodities subject to laws administered by the Bureau of Alcohol, Tobacco, and Firearms.

CI (Commission Interpretation) – designates those categories that have been excluded by the Commission in the light of legislative history of the definition of "consumer commodity." By applying this definition to individual commodities, the Commission has more narrowly applied the latter term and set forth a list of items that do not meet the criteria of consumer commodities. On occasion the Commission is requested in both a formal and informal manner to consider individual products and to determine their status relative to the definition of "consumer commodity" as it is used in the Act.

EPA – designates commodities subject to the Federal Environmental Pest Control Act of 1972 administered by the Environmental Protection Agency.

FDA – designates those commodities which are subject to regulation by the FDA either under the portion of the FPLA administered by that agency or the Federal Food, Drug, and Cosmetic Act (Section 10(a)(3) and Section 7 of the FPLA). Following the code FDA will be a letter further designating the commodity as either a food (F), drug (D), cosmetic (C), or device (DV).

USDA – designates those commodities excluded from jurisdiction by Section 10(a) of the FPLA and represents a commodity within one of the following categories: meat or meat products, poultry or poultry products, or tobacco or tobacco products.

It may be of some help in ascertaining whether a particular product is or is not included within the FPLA definition of "consumer commodity" and thus subject to FTC jurisdiction under that Act, to refer to the following definition:

> ". . . Any article, product, or commodity of any kind or class which is customarily produced or distributed for sale through retail sales agencies or instrumentalities for consumption by individuals, or use by individuals for purposes of personal care or in the performance of services ordinarily rendered within the household, and which is usually consumed or expended in the course of such use."

By applying these criteria to the particular product in question and then reviewing the list of excluded commodities, the observer will be able, in most instances, to determine the status of the item. In the event, however, that the observer is unable to ascertain whether a particular commodity is covered or excluded from FTC jurisdiction, contact FTC for an opinion.

Commodities Included Under FTC Jurisdiction	
Soaps and Detergents	Powder, flakes, chips, etc. Liquid Paste, cake, or tablet
Cleaning Compounds	Liquid Powder Paste or cake Solvent and cleaning fluids for home use
Laundry Supplies	Conditioners and softeners, ironing aids, distilled water Sizings and starches Bluings and bleaches Pre-soaks, enzymes, etc.
Cleaning Devices	Sponges and chamois Steel wool, scouring and soap pads
Food Wraps	Plastic and cellophane Wax paper and paper Foil (aluminum wrap)
Paper Products	Toweling Napkins, table cloths, and place mats Facial tissues Bathroom tissues Disposable diapers Crepe paper Other (e.g., shelf paper, wrapping paper, eye glass tissues)

Commodities Included Under FTC Jurisdiction	
Waxes and Polishes	Powder Liquid Paste and cake Other (e.g., polish impregnated cloths, scratch removers)
Household Supplies	Matches Candles Toothpicks Cordage (string, twine, rope, clothes line, etc.) Drinking straws Lighter and propane torch fuel, flints, pipe cleaners, etc. Lubricants
Household Supplies (continued)	Picnic supplies Sand paper and emory paper Charcoal briquets, chips, logs, etc. Dyes and tints Camera film, photo supplies and chemicals Protective foil cooking utensils Aluminum foil cooking utensils Christmas decorations Solder LPG for other than home heating or cooking Waxes for home use Light bulbs Dry cell batteries Pressure sensitive tapes, excluding gift tapes
Containers	Paper (plain, waxed, or plastic coated) Foil Plastic or Styrofoam
Air Fresheners and Deodorizers	Potpourri
Adhesives and Sealants	
Cordage	

Commodities Excluded from FTC Jurisdiction		
Term	**Description**	**FTC Jurisdiction**
Adhesive Tape		FDA-D
Alcoholic Beverages		BATF
Aluminum Clothesline	Plastic clothesline with a steel core	CI
Antifreeze		CI
Artificial Flowers and Parts		CI
Automotive Accessories	Floor mats, seat covers, spare parts, etc.	CI
Automotive Chemical Products	Auto polish, wax, and finish conditioner, rubbing compound, tire paint, chrome polish, gasoline additives, etc.	CI
Bath Oil and Bubble Bath		FDA-C
Bicycle Tires and Tubes		CI
Books		CI
Bottled Gas	Cooking or heating	CI
Brushes	Bristle, nylon, etc., including hair-brushes, toothbrushes, hand and nail brushes, paint brushes, etc.	CI
Brooms and Mops	Glass, floor, and dish mops, etc.	CI
"Bug Proof" Shelf Paper		EPA
Candle Holders	Without candles	CI
Cameras		CI
Chinaware		CI
Christmas Light Sets	Replacement or other bulbs sold separately are not excluded	CI
Cigarette Lighters		CI
Clothespins		CI
Clothing and Wearing Apparel	Socks, gloves, shoelaces, underwear, etc.	CI
Compacts and Mirrors		CI
Cosmetics	Defined by Section 201(i) of the Food, Drug, and Cosmetic Act as "(l) articles intended to be rubbed, poured, sprinkled, or sprayed on, introduced into, or otherwise applied to the human body or any part thereof for cleansing, beautifying, promoting attractiveness, or altering the appearance, and (2) articles intended for use as a component of any such articles; except that such term shall not include soap."	FDA-C
Cotton Puffs	Sterilized	FDA-D
Crystalware		CI
Detergent Bar with Any Drug or Cosmetic Claim	If the observer experiences difficulty in ascertaining whether or not a given product is a soap or a detergent, contact the manufacturer or FDA.	FDA-D or C
Decorative Magnets		CI
Devices	Defined by Section 201(h) of the Food, Drug, and Cosmetic Act as	FDA-DV

Commodities Excluded from FTC Jurisdiction		
Term	**Description**	**FTC Jurisdiction**
	"instruments, apparatus, and contrivances, including their components, parts, and accessories, intended (1) for use in the diagnosis, cure, mitigation, treatment, or prevention of disease in man or other animals; or (2) to affect the structure or any function of the body of man or other animals." This category includes trusses, syringes, arch supports, etc.	
Diaries and Calendars		CI
Disinfectants		EPA
Drugs	Defined by Section 201(g) (1) of the Food, Drug, and Cosmetic Act as "(a) articles recognized in the official United States Pharmacopeia, official Homeopathic Pharmacopeia, or official National Formulary, or any supplement to any of them; and (b) articles intended for use in the diagnosis, cure, mitigation, treatment, or prevention of disease in man or other animals; and (c) articles (other than food) intended to affect the structure or any function of the body of man or other animals; and (d) articles intended for use as a component of any articles specified in clause (a), (b) or (c); but does not include devices or their components, parts, or accessories."	FDA-D
Durable Articles or Commodities		CI
EPA Covered Products	Products subject to regulation under the Federal Environmental Pesticide Control Act that is administered by the Environmental Protection Agency.	EPA
Fingernail Files		CI
Flowers, Flower Seeds, Fertilizer, and Fertilizer Materials, Plants or Shrubs, Garden and Lawn Supplies		CI
Food	Defined by Section 201(f) of the Food, Drug, and Cosmetic Act as "(1) articles used for food and drink for man or other animals, (2) chewing gum, and (3) articles used for components of any such article."	FDA-F
Fountain Pens, Mechanical Pencils, and Kindred Products	Ball point pens, lead pencils, and lead refills, etc.	CI
Garden Tools	Hoses, trowels, grass clippers, etc.	CI
Germ Killing or Germ Proofing Products		EPA
Gift Tape and Ties	Ribbon, tape, etc.	CI
Gift Wrapping Material	Decorative wrapping foil, paper, cellophane, etc.	CI
Glasses and Glassware	Disposable plastic glasses are not excluded	CI
Gloves (of Any Type)		CI
Greeting Cards		CI
Hair Combs, Nets, and		FDA-DV

Commodities Excluded from FTC Jurisdiction		
Term	**Description**	**FTC Jurisdiction**
Pins		
Hand Tools		CI
Handicraft and Sewing Thread	Yarn, etc.	CI
Hardware	Extension cords, thumb-tacks, hose clamps, nails, screws, picture hangers, etc.	CI
Household Appliances, Equipment, or Furnishings, Including Feather and Down-Filled Products, Synthetic-Filled Bed Pillows, Mattress Pads and Patchwork Quilts, Comforters, and Decorative Curtains		CI
Ink		CI
Insecticides	Insect repellents in any form, mothballs, etc.	EPA
Ironing Board Covers		CI
Jewelry		CI
Lambs Wool Dusters		CI
Luggage		CI
Magnetic Recording Tape	Reels, cassettes, and cartridges.	CI
Meat and Meat Products		USDA
Metal Pails		CI
Motor Oil	Including additives. Household multi-purpose oil is not excluded.	CI
Mouse and Rat Traps		CI
Mouthwash		FDA-D
Musical Instruments		CI
Paints and Kindred Products	Wallpaper, turpentine, putty, paint removers, caulking and glazing compounds, wood fillers, etc. Note, however, that bathroom caulking materials, patching plaster, spackling compound, and plastic wood are not excluded. In the event of uncertainty, contact FTC.	CI
Paintings and Wall Plaques		CI
Pet Care Supplies		CI
Pewterware		CI
Photo Albums		CI
Pictures		CI
Plastic Buckets and		CI

Commodities Excluded from FTC Jurisdiction

Term	Description	FTC Jurisdiction
Garbage Cans		
Plastic Tablecloths, Plastic Place Mats		CI
Plastic Shelf Lining		CI
Pre-Moistened Towelettes		FDA-C
Polishing Cloths	Polishing cloths that are impregnated with polish or chemicals (silicone, etc.) are not excluded.	CI
Poultry and Poultry Products		USDA
Rubber Gloves		CI
Rubbing Alcohol		FDA-D
Safety Flares		CI
Safety Pins		CI
Sanitary Napkins		FDA-D or C
School Supplies	Rulers, crayons, paper, pencils, etc.	CI
Self-Stick Protective Felt Tabs		CI
Seeds of All Kinds		USDA
Sewing Accessories	Needles of any type, thimbles, kindred articles, etc.	CI
Shampoo		FDA-C or D
Shoelaces		CI
Small Arms Ammunition		CI
Silverware, Stainless Steelware, and Pewterware		CI
Smoking Pipes		CI
Soap Bars with a Drug Claim	Including any claim for removing facial blemishes, etc. Refer to Detergent Bars for further discussion in this area.	FDA-D
Soap Dishes		CI
Souvenirs		CI
Sporting Goods		CI
Stationery and Writing Supplies	Looseleaf binders, paper tablets, etc.	CI
Textiles and Items of Wearing Apparel	Cloth laundry bags, towels, cheese cloth, shoe shine cloths, etc.	CI
Tobacco and Tobacco Products	Pipes, cigarettes, etc.	BATF - USDA

Commodities Excluded from FTC Jurisdiction		
Term	**Description**	**FTC Jurisdiction**
Toothpaste		FDA-D
Toys		CI
Typewriter Ribbon		CI
Wire of Any Type		CI
Woodenware		CI

2.6.9. Size Descriptors for Raw, Shell-On Shrimp Products.
(L&R, 1995, p. 97)

Guideline

If size descriptor terms for shrimp (e.g., small, medium, large, or colossal) are used on packages, advertisements, or on signs when offering shrimp for sale from bulk, a statement of count-per-kilogram, if sold by kilogram, or count-per-pound, if sold by pound, should be included adjacent to the size descriptor (e.g., medium-large, 31 to 40 shrimp per pound).

2.6.10. Model Guidelines for the Administrative Review Process.

Purpose

These guidelines are provided to assist weights and measures programs in establishing an administrative review process. They are not intended to be the only process an agency may use nor are they intended to supersede any agency's existing process. Before implementing ANY process, it should be approved by legal counsel.

These guidelines ensure that persons affected by "inspection findings" (e.g., price misrepresentations or shortweight packages), or who are deprived of the use of their property (devices or packages placed under "stop" or "off-sale" order), are provided a timely-independent review of the action. The process enables affected persons to provide evidence which could be relevant in determining whether the enforcement action was proper. The purpose of the process is to ensure that a person's ability to conduct business is not hindered by improper enforcement actions. This process is independent of any other action (e.g., administrative penalties, prosecutions, etc.) that may be taken by the enforcement agency.

Background

In the course of their work, weights and measures officials take enforcement actions that may prohibit the use of devices or the sale of packaged goods (e.g., "stop-sale" or "off-sale" orders for packages and "stop-use" or "condemnation" tags issued on devices). Improper actions (e.g., not following prescribed test procedures, enforcing labeling requirements on exempted packages, or incorrectly citing someone for a "violation") place the official and the jurisdiction in the position of being liable for the action if it is found that the action was "illegal." In some cases, weights and measures jurisdictions could be ordered to pay monetary damages to compensate the affected party for the improper action.

This process is one way to provide affected persons an opportunity to present evidence which may be relevant in determining whether the order or finding has been properly made to an independent party. The procedure enables business operators to obtain an independent review of orders or findings so that actions affecting their business can be evaluated administratively instead of through litigation. This ensures timely review, which is essential because of the impact that such actions may have on the ability of a business to operate and in cases where perishable products may be lost.

Review Provisions

Parties affected by enforcement actions must be given the opportunity to appeal enforcement actions.

Inspectors are the primary contacts with regulated firms and thus are in the best position to ensure that the enforcement actions they take are "proper." "Proper" means that inspections are conducted (1) within the scope of the authority granted by law, (2) according to recognized investigative or testing procedures and standards, and (3) that enforcement actions are lawful. The "burden" for proving that actions are "proper" falls on the weights and measures program, not on regulated firms.

Weights and measures officials are law enforcement officers. Therefore, they have the responsibility to exercise their authority within the "due process" provisions of the U.S. Constitution. As weights and measure programs carry-out their enforcement responsibilities in the future, more and more challenges to their actions and authority will occur. It is in the best interest of any program to establish strict operational procedures and standards of conduct to prevent the occurrence of improper actions which may place the jurisdiction in an untenable position in a court challenge of an enforcement action. The foundation for ensuring "proper" actions is training, clear and concise requirements, and adoption of, and adherence to uniform test procedures and legal procedures.

Prior to taking enforcement actions the inspector should recheck test results and determine that the information on which the action will be taken is accurate.

Inspections shall be conducted with the understanding that the findings will be clearly and plainly documented and reviewed with the store's representative.

During the review of the findings, the firm's representative may provide information which must be used by the inspector to resolve the problems and concerns before enforcement actions are taken. In some cases, the provided information may not persuade the inspector to forego the action. In some cases, the inspector and business representative may not understand the circumstances surrounding the violations, or there may be a conflict between the parties that they cannot resolve. In other cases, the owner or manufacturer may not learn that an enforcement action has occurred until long after the inspector leaves the establishment.

Steps:

1. Provide a framework that will help in resolving most of these situations where "due process" is of concern. Make sure that the responsible party (e.g., as declared on the package label) is notified of violations and receives copies of inspection reports. Establish standard operating procedures to assure the affected party of timely access to a representative of the weights and measures program so that the firm can provide the relevant information or obtain clarification of legal requirements.

2. Make the process as simple and convenient as possible. Especially in distant or rural areas where there are no local offices, the review should be conducted by a supervisor of the official taking the action if agreed to by the person filing the request for review.

3. The process should include notice that the firm can seek review at a higher level in the weights and measures program or an independent review by a third party. The following procedures are recommended:

   (a) Any owner, distributor, packager, or retailer of a device ordered out of service, or item or commodity ordered "off-sale," or inspection finding (e.g., a price misrepresentation or a shortweight lot of packages) shall be entitled to a timely review of such order, to a prompt, impartial, administrative review of such off-sale order or finding.

   A notice of the right to administrative review should be included on all orders or reports of findings or violations and should be communicated to the responsible firm (e.g., person or firm identified on the product label):

> Sample Notice
>
> You have the right to Administrative Review of this order or finding. To obtain a review, contact the Director of Weights and Measures by telephone or send a written request (either postmarked, faxed, or hand delivered) to:
>
> (Name, Address or Fax Number of the Director or other Designated Official)
>
> Your request should reference any information that you believe supports the withdrawal or modification of the order or finding.

(b) The administrative review shall be conducted by an independent party designated by the Director or before an independent hearing officer appointed by the Department. The officer shall not be a person responsible for weights and measures administration or enforcement.

(c) No fees should be imposed for the administrative review process.

(d) The firm responsible for the product or the retailer may introduce any record or other relevant evidence.

For example:

i. Commodities subject to the off-sale action or other findings were produced, processed, packaged, priced, or labeled in accordance with applicable laws, regulations or requirements.

ii. Devices subject to the "stop-use" order or "condemnation" were maintained in accordance with applicable laws, regulations or requirements.

iii. Prescribed test procedures or sampling plans were not followed by the inspector.

iv. Mitigating circumstances existed which should be considered.

(e) The reviewer must consider the inspector's report, findings, and actions as well as any evidence introduced by the owner, distributor, packager, or retailer as part of the review process.

(f) The reviewer must provide a timely written recommendation following review unless additional time is agreed to by the department and the petitioner.

(g) The reviewer may recommend to the Department that an order be upheld, withdrawn or modified. If justified the reviewer may recommend other action including a reinspection of the device or commodity based upon information presented during the review.

(h) All actions should be documented and all parties advised in writing of the results of the review. The report of action should be detailed in that it provides the reasons for the decision.

2.6.11. Good Quantity Control Practices.

Good Quantity Control Practices means that the plant managers should take all reasonable precautions to ensure the following quantity control standards or their equivalent are met:

1. A formal quantity control function is in place with authority to review production processes and records, investigate possible errors, and approve, control, or reject lots.

2. Adequate facilities (e.g., equipment, standards and work areas) for conducting quantity control functions are provided and maintained.

3. A quantity control program (e.g., a system of statistical process control) is in place and maintained.

4. Sampling is conducted at a frequency appropriate to the product process to ensure that the data obtained is representative of the production lot.

5. Production records are maintained to provide a history of the filling and net content labeling of the product.

6. Each "production lot" contains on the average the labeled quantity and the number of packages exceeding the specified maximum allowable variation (MAV) value in the inspection sample shall be no more than permitted in Tables 2-1. Class of Scale and 2-2. Acceptance Tolerances for Class of Scale Based on Test Load Divisions in NIST Handbook 133.

7. Packaging practices are appropriate for specific products and measurement procedures (e.g., quantity sampling, density and tare determinations) and guidelines for recording and maintaining test results are documented.

8. Personnel responsible for quantity control follow written work instructions and are competent to perform their duties (e.g., background, education, experience and training). Training is conducted at sufficient intervals to ensure good practices.

9. Recognized procedures are used for the selection, maintenance, adjustment, and testing of filling equipment to insure proper fill control.

10. Weighing and measuring devices are suitable for their intended purpose, and measurement standards are suitable and traceable to national standards. This includes a system of equipment maintenance and calibration to include recordkeeping procedures.

11. Controls over automated data systems and software used in quantity control ensure that information is accessible, but changeable only by authorized personnel.

12. Tare materials are monitored for variation. Label changes are controlled to ensure net quantity matches labeled declaration.

2.6.12. Point-of-Pack Inspection Guidelines.

A. Weights and Measures Officials' Responsibilities.

1. Conduct inspections during hours when the plant is normally open for business. Open the inspection by making contact with the plant manager or authorized representative (e.g., the quality assurance manager or the production manager).

2. Present the proper credentials and explain the reason for the visit (e.g., routine or follow-up inspection or consumer complaint).

3. Request access to quantity measurement equipment in the packing room, moisture testing equipment in the laboratory or in the packing room, and product packed on premise or stored in warehouse areas.

4. Obtain permission from a plant representative prior to using a tape recorder or a camera.

5. Conduct inspection related activities in a professional and appropriate manner and, if possible, work in an area that will not interfere with normal activities of the establishment.

6. Abide by all the safety and sanitary requirements of the establishment and clean the work area upon completion of the inspection/test. Return borrowed equipment and materials.

7. To close the inspection, recheck inspection reports in detail and ascertain that all information is complete and correct.

8. Sample questions and tasks for Inspectors:

   (a) Inside Buildings and Equipment.

      i. Is all filling and associated equipment in good repair?

      ii. Are net content measurement devices suitable for the purpose being used?

      iii. Are standards used by the firm to verify device accuracy traceable to NIST?

   (b) Packing Room Inspection.

      i. Observe if the program for net quantity of content control in the packing room is actually being carried out.

      ii. Ensure that the weighing systems are suitable and tare determination procedures are adequate. If there are questions regarding tare determination, weigh a representative number of tare and/or filled packages.

      iii. For products labeled and filled by volume and then checked by weight, ensure that proper density is used.

   (c) Warehouse Inspection.

   If an inspection is conducted:

      i. Select lot(s) to be evaluated.

      ii. Determine the number of samples to be inspected. Use the appropriate sampling plan as described in NIST Handbook 133, "Checking the Net Content of Packaged Goods."

      iii. Randomly select the number of samples or use a mutually agreed on plan for selecting the samples.

      iv. Determine the average net quantity of the sample and use the standard deviation factor to compute the Sample Error Limit (SEL) to evaluate the lot.

      v. Look for individual values that exceed the applicable Maximum Allowable Variation as found in NIST Handbook 133.

      vi. Apply moisture allowances, if applicable.

      vii. Review the general condition of the warehouse relevant to package integrity, good quantity control, and distribution practices.

      viii. Prepare an inspection report to detail findings and actions.

9. Close the inspection - Review findings with Plant Representative.

After the inspection, meet with the management representative to discuss inspection findings and observations. Provide additional information as needed (e.g., information on laws and regulations or explanations of test procedures used in the inspection). Be informative, courteous and responsive. If problems/violations are found during the inspection/test, bring them to the attention of the appropriate person.

B. Plant Management Responsibilities.

1. Recognize that inspectors are enforcing a federal, state or local law.

2. Assist the official in conducting inspection activities in a timely and efficient manner.

3. During the initial conference with the inspector, find out whether the inspection is routine, a follow-up, or the result of a consumer complaint. If a complaint, obtain as much information as possible concerning the nature of the complaint, allowing for an appropriate response.

4. The plant manager, quality assurance manager, or any designated representative should accompany the inspector.

5. Plant personnel should take note of the inspector's comments during the inspection and prepare a detailed write-up as soon as the inspection is completed.

6. When an official presents an inspection report, discuss the observations and, if possible, provide explanations for any changes deemed necessary as a result of the inspection/test.

Plant Management: information that must be shared with the inspector.

1. Establishment name and address.

2. Type of firm and information on related firms or applicable information (e.g., sub-contractor, servant, or agent).

3. General description and location of shipping and storage areas where packaged goods intended for distribution are stored.

4. Commodities manufactured by or stored at the facility.

5. Names of responsible plant officials.

Plant Management: information that may be shared with the inspector.

1. Simple flow sheet of the filling process with appropriate net content control checkpoints.

2. Weighing or measuring device maintenance and calibration test records.

3. Type of quantity control tests and methods used.

4. Net content control charts for any lot, shipment, or delivery in question or lots which have previously been cited.

5. Method of date coding the product to include code interpretation.

6. Laboratory reports showing the moisture analysis of the products which are in question or have been previously cited.

7. Product volume of lot sizes or related information.

8. Distribution records related to a problem lots including names of customers.

2.6.13. Guideline for Verifying the Labeled Basis Weight of Communication and Other Paper.
(L&R, 1998, p. 27)

2.6.13.1. Equipment. – Linear measure recommended in Section 5.3.1. Equipment in the third edition of NIST Handbook 133 "Checking the Net Contents of Packaged Goods."

- Scale with a minimum division of 0.5 g (0.001 lb) or less.
- Scientific calculator with a sample standard deviation function.

2.6.13.2. Scope and Recommended Enforcement Approach. – Paper is manufactured in various "basis weights" for use in different applications (e.g., copy paper can have a basis weight of 18 or 20 lb). Basis weight is part of the product identity and not a declaration of net contents. This procedure is used to audit the basis weight declared on package labels. If the tested packages in a sample do not have an average basis weight equal to or greater than the labeled basis weight, the inspection lot may be in violation. A potentially violative lot should be placed "off-sale" until the owner provides documentation to confirm that the labeled basis weight corresponds to the basis weight declared by the original manufacturer. If documentation is not provided, the inspection lot should remain "off-sale" until the basis weight declaration is corrected.

2.6.13.3. Determine Target Net Weight for Common Types of Paper. – The basis weight of paper is the designated weight (measured in grams or pounds per specified area) of one ream in basic sheet size for the type of paper being tested. This procedure permits the confirmation of basis weight by linear measurement and gravimetric testing. This procedure is designed to test the various types, size, count, and basis weights of packaged paper currently in the marketplace. Table 1 lists the "area of basic sheet size" for common types of paper. A "ream" equals 500 sheets of basic sheet size for all types of paper other than tissue paper. A "ream" of tissue paper equals 480 sheets. Each of the standard categories of paper products shown in Table 1 has a different standard basic sheet size. Although there are basic sheet sizes, paper is packaged and marketed in various sizes and counts. The net weight of packaged paper can be determined from the label information using the General Formula for Sheet Paper. For roll paper, use one (1) for the sheet count.

General Formula for Sheet Paper

$$\frac{PA \; x \; BW}{BSS} \; x \; \frac{SC}{500} = TNW$$

Where:

PA	=	measured area of one sheet of paper
BW	=	labeled basis weight
BSS	=	area of basic sheet size from Table 1
SC	=	labeled package sheet count
TNW	=	target net weight of paper

2.6.13.4. Test Procedure. – The following gravimetric, measuring, and counting procedures shall be used to determine if packages are accurately labeled. Procedures are also provided for verifying net quantity of content declarations for count and dimensions (e.g., length and width.)

2.6.13.4.1. Sample Selection. – Select a sample from an inspection lot using Table 2-1 Sampling Plans of Category A (page A-2) in the fourth edition of NIST Handbook 133, "Checking the Net Contents of Packaged Goods." Determine an average tare weight in accordance with Section 2 of the fourth edition of NIST Handbook 133.

2.6.13.4.2. Determine Target Net Weight of Common Types of Paper Packaged in Various Sizes or Counts.

Verify the basis weight declared on a package using the following gravimetric procedure:

a. Record the following information from the package label on a worksheet. (See Figure 1 for a sample label.)

   1. Type of Paper (TP)
   2. Length (L)
   3. Width (W)
   4. Package Sheet Count (PSC)
   5. Basis Weight (BW)
   6. Basic Size Sheet (BSS)

b. Compute the Target Net Weight (TNW) for the sample packages using the General Formula for Sheet Paper. TNW is what the paper should weigh if the labeled properties of the packaged paper are accurate.)

c. Determine the average net weight of the sample packages. (Do not use sample error limit calculations.) If the average net weight is not equal to or more than the Target Net Weight, go to Section 2.6.14.3. to determine if the labeled basis weight (BW) is correct. If the average net weight is equal to or more than the labeled basis weight, the sample passes.

Basis Weight Worksheet (see Figure 1)

Type of Paper (TP):	Copy Paper
Length (L):	11 in
Width (W):	8½ in
Area (PA) of Sheet (L×W):	93.5 in^2
Package Sheet Count (PSC):	500
Basis Weight (BW):	20 lb
Basic Sheet Size (BSS):	17 in × 22 in
Area of BSS from Table 1 or by calculation:	374 in^2

Use the General Formula to compute Target Net Weight (TNW):

Target Net Weight (TNW) = 5 lb

Example
White Copy Paper
75 g/m^2 (20 lb) Bond

Size: 216 mm × 279 mm (8½ in × 11 in)

Count: 500 Sheets

Figure 1. Sample Label

NOTE: Three factors will cause actual sample weights to differ from the TNW:
Actual sheet count in package
Actual basis weight of paper being tested
Actual dimensions of the paper being tested

$$\frac{(93.5\,in^2 \times 20\,lb)}{374\,in^2} x \frac{500}{500} = 5\,lb$$

2.6.13.4.3. Determine Basis Weight.

This procedure is used to identify potentially violative packages. If the Average Basis Weight (ABW) for the sample determined by this procedure is not equal to or greater than the labeled basis weight, other steps must be taken. Moisture affects the weight of paper, but the moisture content of paper can only be determined in a measurement laboratory according to the Technical Association of the Pulp and Paper Industry (TAPPI) (URL: **www.tappi.org/**) TAPPI – T410 om-08, "Grammage of Paper and Paperboard (Weight per Unit Area)."

a. Verify the basis weight for each package according to the following steps:

i. Identify the paper type from Column 1 in Table 1 and record the area for the paper type from Column 2.

ii. Select a sample of paper from each of the tare sample packages. Use a sample of exact count to eliminate the possibility that the packages are short count.

- For packages with more than 100 sheets, use 100 sheets; or
- For packages with 100 sheets or less, verify the sheet count and use all of the sheets.

iii. Use a basis weight work sheet and determine the number of basic size sheets the paper sample represents with the following formula:

$$\frac{PA}{A} \; x \; EC \; = \; ENBSS$$

Where:

A	=	area of basic sheet size from Table 1
PA	=	area (l x w) of one sheet of paper
EC	=	exact sheet count of sample
ENBSS	=	equivalent number of basic size sheets

iv. Determine the average basis weight,

Where:

BW	=	basis weight for each package
ABW	=	average basis weight
ENBSS	=	equivalent number of basic size sheets from step iii
NW	=	net weight of sample
RC	=	Ream Count (500; for tissue paper, use 480)

v. Repeat this step for each paper package from the tare sample and average the basis weights to obtain an Average Basis Weight (ABW). If the ABW is less than the labeled basis weight, or if the difference between the basis weight of the sample packages is more than 1 scale division, measure and compute the basis weight for each of the remaining packages.

vi. Weigh each sample. If the basis weight from step iv is less than the labeled basis weight, re-calculate the target net weight by using the general formula for sheet paper.

Table 1. Common Types of Paper and Area of Basic Sheet Size	
Paper Type	**Area**
Bond, Ledger, Thin, Writing, and Track Feed Printer Paper	2412 cm^2 (374 in^2)
Manuscript Cover	3599 cm^2 (558 in^2)
Blotting	2941 cm^2 (456 in^2)
Cover	3354 cm^2 (520 in^2)
Blanks	3974 cm^2 (616 in^2)
Printing Bristols	4135 cm^2 (641 in^2)
Wrapping, Tissue, Waxed, Newsprint and Tag Stock	5574 cm^2 (864 in^2)
Book, Offset, and Text	6129 cm^2 (950 in^2)
Index Bristol	5019 cm^2 (778 in^2)

vii. Use the target net weight computed in step vi and re-weigh the inspection lot samples using the Section 2. of the fourth edition of NIST Handbook 133. If inspection sample weights differ from the target net weight computed using the average basis weight determined in vi, the label sheet count is probably inaccurate.

b. Verify the label sheet count by counting the number of sheets in each package.

c. Verify sheet dimensions (length x width) for each package of the sample.

$$\frac{NW \ x \ RC}{ENBSS} = BW$$

2.6.13.4.3.1. Other Types of Packaged Paper.

1. Roll Paper. – When testing rolled paper, cut a length of paper from the roll equal to 9350 divided by the width of the paper in inches. Make sure the ends of this length of paper are square. Proceed to section 2.6.14.3 step a. Disregard the exact sheet count in step iii.

2. Continuous Track Feed Printer Paper:

   i. Count out a sample of 100 sheets from each tare sample package of the inspection lot.

   ii. Weigh each 100 sheet sample and record the weights.

   iii. Calculate an average weight.

   iv. Remove printer track feed strips from each sample.

   v. Re-weigh each sample after the tractor feed has been removed and record the weights.

   vi. Calculate an average weight from step v.

   vii. Calculate percentage (%) difference in the average weights in steps iii and vi.

viii. After the track feed strips have been removed, use the samples to verify the basis weight for the packages of the inspection lot using the formula in 2.6.14.2. If the basis weight is less than the labeled basis weight, refer to 2.6.13.2.

ix. If the basis weight established in step viii is the same as the labeled basis weight, weigh the remaining packages from the sample and compare the actual net weights with the TNW. (Remember to adjust the TNW up by the percentage established in step vii.)

x. If the adjusted weights of the remaining samples is less than the TNW, the deficiency may have been caused by:

a. the sheet count in the package.

b. the basis weight of the paper.

c. the dimensions of the paper.

d. combinations of the above.

This procedure is for use in verifying that the basis weight included in a statement of identity is not misleading or deceptive. It is not intended to be used as the final criterion on which enforcement action is taken. Instead, the test procedure is only used to identify potentially volatile lots. There are two alternative actions that can be taken if the test results indicate that a lot is potentially volatile. The first is to review the documentation supplied by the original manufacturer to the converter to determine if any misrepresentation has occurred. The second is to collect packages of the paper and test them according to the latest version of ASTM International Method D646 for "Grammage of Paper and Paperboard."

2.6.14. Labeling Guidelines for Chamois.

(L&R, 1999, p. L&R-25)

These requirements are based on the Uniform Packaging and Labeling Regulation in the 1999 edition of NIST Handbook 130, "Uniform Laws and Regulations" and regulations and guidelines of the Federal Trade Commission.

General

The following information must be declared on the principal display panel of the chamois package. The principal display panel is the tag, or label that consumers can examine under normal and customary conditions of display.

- Identity - what the package contains
- Net Quantity of Contents - how many items the package contains and the area of the item(s)

The following information may appear anywhere on the package.

- Responsibility – the party responsible for packaging or distributing the product.

2.6.14.1. Declaration of Identity. – Chamois is a natural product made of sheepskin which has been oil tanned. In 1964, the FTC issued an advisory opinion stating that using the word "chamois" on a product (e.g., "Artificial" Chamois, "Synthetic" Chamois, "Pig Chamois" or "Man Made" Chamois) that is not made from oil tanned sheepskin is unlawful and deceptive. Packages are required to declare identity in terms of:

i. the name specified in or required by any applicable federal or state law or regulation or, in the absence of this,

ii. the common or usual name or, in the absence of this,

iii. the generic name or other appropriate description, including a statement of function.

Example:
Chamois, Natural Chamois Leather

2.6.14.2. Declaration of Net Quantity of Contents. – The following information is required to appear on the lower 30 % of the principal display panel of all packages:

Count

- The package must include a count declaration (e.g., 1 Chamois) unless the statement of identity clearly expresses the fact that only one unit is contained in the package. A package containing two or more units shall bear a statement in terms of count (e.g., 2 Chamois).

Area

- Chamois packages must have area declarations in both inch-pound and metric units.

Metric

- For areas that measure less than 1 m^2, the area shall be stated in square decimeters and decimal fractions of a square decimeter or in square centimeters and decimal fractions of a square centimeter;
- For areas that measure 1 m^2 or more, the area shall be stated in square meters and decimal fractions to not more than three places.

To facilitate value comparison and simplify the measurement process, chamois should be measured in one quarter square foot (2.322 57 decimeter) increments. Dimensions should be rounded down to avoid overstating the area.

For example: 2 square feet (18.5 square decimeters) or 2 ft^2 (18.5 dm^2)

Conversion Factors:

1 ft^2 = 9.290 30 dm^2

1 in^2 = 6.451 6 cm^2

1 yd^2 = 83.612 7 dm^2

Inch-pound Units

- For areas that are less than 1 ft^2 (929 cm^2), the area declaration shall be expressed in square inches and fractions of square inches;
- For areas of 1 ft^2 (929 cm^2), or more, but less than 4 ft^2 (37.1 dm^2), the area shall be expressed in square feet with any remainder expressed in square inches or in fractions of a square foot;
- For areas of 4 ft^2 (37.1 dm^2) or more, the area should be expressed in terms of the largest whole unit (e.g., square yards, square yards and square feet, or square feet) with any remainder expressed in square inches and fractions of a square inch or in fractions of the square foot or square yard.

Chamois labeled for retail sale is exempt from these requirements if (a) the area of a full skin is expressed in terms of square feet with any remainder in terms of the common or decimal fraction of the square foot

(929 cm^2), or (b) the area for cut skins of any configuration is expressed in terms of square inches and fractions thereof. Where the area of a cut skin is at least one square foot (929 cm^2) or more, the statement of square inches shall be followed in parentheses by a declaration in square feet with any remainder in terms of square inches or common or decimal fractions of the square foot.

Prohibited Labeling Practices

- Do not use qualifying terms or phrases (e.g., "Approximate Size," "Size when Wet," "Up to 20 % Larger When Wet").
- Do not use unacceptable symbols (e.g., using (") as a symbol for inches is not acceptable).

2.6.14.3. Declaration of Responsibility. – The name and address of the manufacturer, packer, or distributor must be conspicuously specified on the label of any package that is kept, offered, exposed for sale, or sold anywhere other than the premises where packed. The name shall be the actual corporate name, or, when not incorporated, the name under which the company does business. This declaration does not have to appear on the principal display panel.

For example:
Chamois Tanning Company
8190 Main Road
Tarpon Springs, FL 34568

The address shall include street address, city, state (or country if outside the United States), and ZIP Code (or the postal code, if any, used in countries other than the United States); however, the street address may be omitted if it is shown in a current city directory or telephone directory.

Sample Labels

1. If one natural chamois is in a see through package, the following label would be acceptable:

Natural Chamois Leather Distributed by: Chamois Leather Co. 8190 Main Road Tarpon Springs, FL 34568 **7 ft^2 (65 dm^2)**

2. The next sample would apply if one chamois is in a package and the statement of identity does not clearly express the fact the package only contains one unit.

Chamois Chamois Leather Company 8190 Main Road Tarpon Springs, FL 34568 One Chamois **3 ft^2 (27.8 dm^2)**

2.6.15. Labeling Guidelines for Natural and Synthetic Sponges.

(L&R, 1999, p. L&R-31)

These requirements are based on the Uniform Packaging and Labeling Regulation in NIST Handbook 130, "Uniform Laws and Regulations" and regulations and guidelines of the Federal Trade Commission. All indicated dimensions and conversions from metric to inch-pound units are approximate only and are used for illustration purposes only.

General

The following information must be declared on the principal display panel (PDP) of a package of sponge(s). The PDP is the part of label (or package) most likely to be displayed, presented, shown to or examined by consumers. A tag or spot label may be used.

- Identity – what the package contains
- Net Quantity of Contents – how many items in the package and the dimensions of the item(s)

The following information may appear anywhere on the package.

- Responsibility – the name of the processor or distributor

2.6.15.1. Declaration of Identity.

a. A declaration of identity that clearly describes the origin and other relevant information about the sponge must appear on the label of each package. The identity of a sponge must include information about its origin (i.e., is it a natural or synthetic sponge). The identity shall be in terms of (i) the name specified in or required by applicable federal or state law or regulation, or (ii) the common or usual name, or (iii) the generic name or other appropriate description.

   For example:

   Sea Wool Sponge, Rock Island Sponge, Sea Grass Sponge, Sea Yellow Sponge, or Atlantic Silk Sponge

   - Origin - Natural or Synthetic
   - For natural sponges, the label must specify if they are "Cut" or "Form." "Cut" sponges are those that have been cut into halves, quarters, or fourths while "forms" are whole sponges.
   - For natural sponges, indicate type of sponge (e.g., "silk," "seawool," or "yellow")

b. Identifiers

   - Terms which indicate locations of origin on some natural sponges (e.g., "Atlantic Sea Sponge") are permitted to be used for identification if they accurately describe the source of the sponge.
   - Use of terms that may be interpreted by consumers to imply quality, durability, or "expert" endorsement (e.g., "professional quality sponge") are permitted as identifiers if they are not misleading. However, terms that imply quality should be used with care if they are not based on a recognized grading system. Use of terms to describe sponge texture such as "fine," "medium," or "coarse" are acceptable.

2.6.15.2. Declaration of Net Quantity of Contents. – The following information must appear on the lower 30 % of the principal display panel of all packages:

- Count

  The package must include a count declaration (e.g., 1 sponge) unless the statement of identity clearly expresses the fact that only one unit is contained in the package. A package containing two or more units shall bear a statement in terms of count (e.g., 2 sponges).

- Dimensions

  The package must include the dimensions of the sponges in inches and centimeters.

Silk Sponges

To facilitate value comparison and simplify the measurement process, sponges should be measured in ½ in (1 cm) increments. Dimensions should be rounded down to avoid overstating the size of a sponge.

For example:
- 6 in, 6½ in, and 7 in for inch declarations;
- 15 cm, 16 cm, and 17 cm for metric declarations

- Synthetic sponges: the dimensions shall include length x width x height (thickness). Either unit of measure can be the primary declaration (e.g., the metric or inch-pound units can be presented first).

  1 Sponge 17 x 10 x 5 cm (7 x 4 x 2 in)

- Natural sponges: the declaration shall be a single measurement representing the maximum dimension of one axis of a sponge that is passed through a circular template. When measured, the sponge is "classified" as a specific size when at least three (including two opposing) points of the sponge touch the template (e.g., see graphic on the following page where the sample sponge is designated as a 7 in [17 cm] sponge).

As the following pictures show, natural sponges are irregular in size and shape and have traditionally been measured using this procedure. It is difficult to develop a meaningful or cost effective measurement process that would provide a means of direct comparison between synthetic and natural sponges based on dimensions. Requiring declarations, such as average height, length, or width of natural sponge procedures would increase the costs for industry and consumers.

Sea Wool Sponges

Sea Grass Sponges

This graphic illustrates an irregular form of a natural sponge passing through a 17 cm (7 in) template and touching at least two opposite points. This sponge could be labeled 7 in.

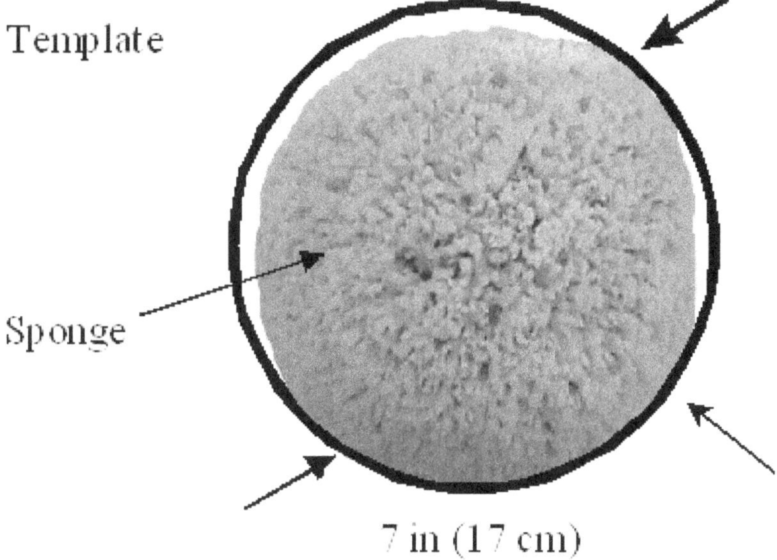

- For banana sponges the size will be determined as shown below. This sponge is 17 cm (7 in).

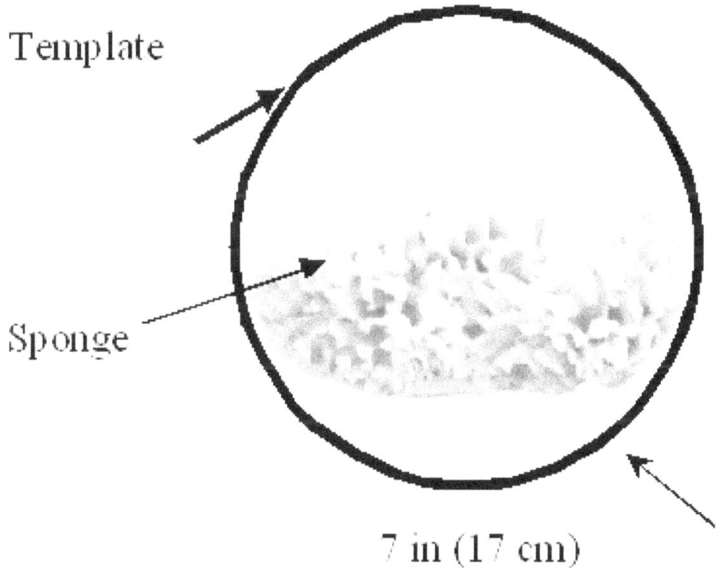

Good Measurement Practice

- Dimensions are determined with the sponge wet.
- Measuring templates (see photo below for the currently used type templates):

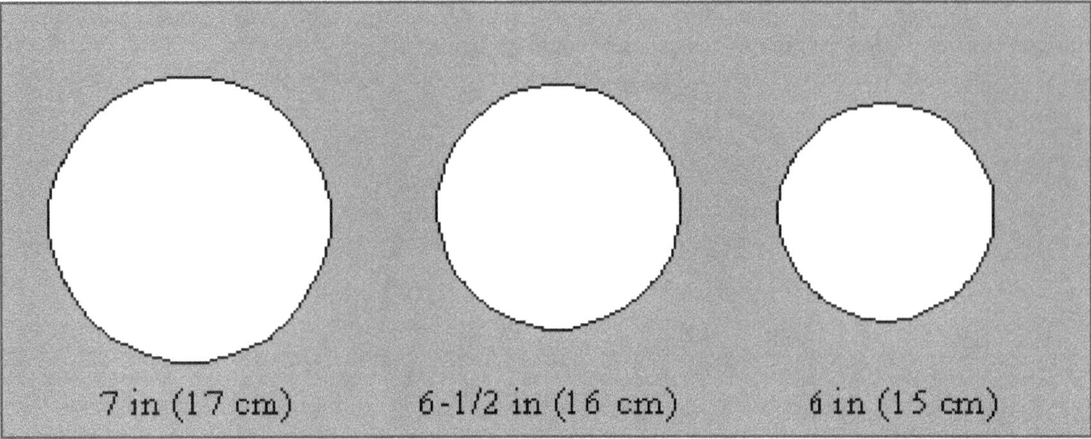

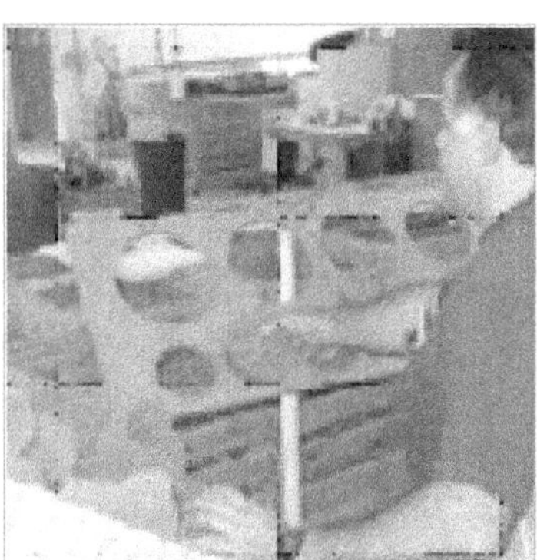

- should be constructed of rigid metal or plastic material.
- circular openings should graduate in increments of one-half inch (one centimeter).
- The error in the circular openings shall not be greater than ± $^1/_{32}$ in (± 0.79 mm) as specified in Table 2. Tolerances in Section 5.52. Linear Measures of NIST Handbook 44 "Specifications, Tolerances, and Technical Requirements for Weighing and Measuring Devices."

Prohibited Labeling Practices

- Stating country of origin declarations that are not accurate.
- Declaring ranges of dimensions (e.g., 4″- 5″ in) or using terms such as "half or semi form" instead of either "cut" or "form."
- Using qualifying terms. (e.g., "Wet Size," "Approximate" or "Jumbo")
- "Anti-bacterial" claims must meet EPA requirements.
- Using type size that does not meet minimum height requirements.
- Using unacceptable symbols (e.g., using (") as a symbol for inches is not acceptable).

2.6.15.3. Declaration of Responsibility. – The name and address of the processor or distributor must be specified on the label of any package that is kept, offered, or exposed for sale, or sold anywhere other than the premises where packed. The name shall be the actual corporate name or, when not incorporated, the name under which the business is conducted.

For example:

Processed by
Argonaut Sponge Company
8190 Main Road
Tarpon Springs, Florida 34568

The address shall include street address, city, state (or country if outside the United States), and ZIP Code (or the postal code, if any, used in countries other than the United States); however, the street address may be omitted if this is shown in a current city directory or telephone directory.

Sample Labels

Yellow Sponge Cut Argonaut Sponge Company 8190 Main Road Tarpon Springs, FL 34568 **One - 17.5 cm (7 in)**

If a natural sponge is in a box, carton, or package that does not permit consumers to see how many sponges are in the box, the package must include a count declaration (e.g., 1 sponge) unless the statement of identity clearly expresses the fact that only one unit is contained in the package. A package containing two or more units shall bear a statement in terms of count (e.g., 2 sponges). The following sample label would apply.

Synthetic Sponge Made by: Argonaut Sponge Company 8190 Main Road Tarpon Springs, FL 34568 **17.7 x 10 x 5 cm (7 x 4 x 2 in)**

Synthetic Sponge Made by: Argonaut Sponge Company 8190 Main Road Tarpon Springs, FL 34568 **1 - Sponge 17.7 cm x 10 x 5 cm (7 in x 4 in x 2 in)**

If a package does not permit the consumer to see how many sponges are the box, it must include a count declaration (e.g., 1 sponge) unless the statement of identity clearly expresses the fact that only one unit is contained in the package. A package containing two or more units shall bear a statement in terms of count (e.g., 2 sponges). A transparent bag of small pieces of sponge may be sold on the basis of count if the words "Irregular Dimensions" appear in conjunction with the declaration of count (e.g., 10 Sponges - Irregular Dimensions).

2.6.16. Minimum Fuel Flush for Octane Verification.
(L&R, 2000, p. L&R-13)

A minimum of 1.2 L (0.3 gal) of motor fuel shall be flushed from a dispenser before taking a sample for octane verification. The flush shall be returned to the storage tank containing the lowest octane.

THIS PAGE INTENTIONALLY LEFT BLANK

Index

A

B

C

D

E

F

G

H

I

J

K

L

M

N

O

P

T

U

V

W

Y

www.ingramcontent.com/pod-product-compliance
Lightning Source LLC
Chambersburg PA
CBHW080236180526
45167CB00006B/2304
9781505262209